Advanced Level

PURE MATHEMATICS

Key Facts

By Nji Emmanuel Ndi

TEL: (+237) 7668 4050

E-mail: manuelndike@gmail.com

First Edition

Printed by CreateSpace, an Amazon.com Company

eStore address: www.CreateSpace.com/5996097

Available from Amazon.com, CreateSpace.com, and other retail outlets

Available on Kindle and other retail outlets

Also by the same author:

Complete Ordinary Level Mathematics Passport.

Rudiments of Ordinary Level Mathematics.

DEDICATION

Dedicated to all mathematics teachers
and students past and present.

Table of Contents

TOPIC 1: QUADRATIC THEORY................................3

TOPIC 2: INEQUALITIES AND INEQUATIONS.............11

TOPIC 3: INDICES, LOGARITHMS AND SURDS 23

TOPIC 4: POLYNOMIALS WITH REAL COEFFICIENTS..28

TOPIC 5: PARTIAL FRACTIONS 30

TOPIC 6: TRIGONOMETRY 31

TOPIC 7: LIMITS AND DIFFERENTIATION 43

TOPIC 8: INTEGRATION 53

TOPIC 9: FIRST ORDER DIFFERENTIAL EQUATIONS...62

TOPIC 10: SEQUENCE AND SERIES 64

TOPIC 11: PERMUTATIONS AND COMBINATIONS 68

TOPIC 12: COORDINATE GEOMETRY........................73

TOPIC 13: COMPLEX NUMBERS 84

TOPIC 14: VECTORS..89

TOPIC 15: RELATIONS...101

TOPIC 16: MAPPINGS AND FUNCTIONS 106

TOPIC 17: NUMERICAL METHODS 122

TOPIC 18: MATRICES AND TRANSFORMATIONS.....126

TOPIC 1
QUADRATIC THEORY

Quadratic Expressions
A quadratic expression is of the form $ax^2 + bx + c$, where a, b and c are constants and $a \neq 0$.

Expansions Leading to Quadratic Expressions
Monomial and Binomial Expansions
$$x(x + b) = x^2 + bx$$
$$x(x - b) = x^2 - bx$$

Binomial Expansions
$$(x + a)(x + b) = x^2 + (a + b)x + ab$$
$$(x - a)(x - b) = x^2 - (a + b)x + ab$$
$$(x - a)(x + b) = x^2 - (a - b)x - ab$$

The Square of a Binomial
$$(x + a)^2 = x^2 + 2ax + a^2$$
$$(x - a)^2 = x^2 - 2ax + a^2$$

The Difference of Two Squares
$$(x + a)(x - a) = x^2 - a^2$$

Factorising a Quadratic Expression
To factorise the quadratic expression $ax^2 + bx + c$,
1. Multiply a by c to have ac,
2. Find a pair of integral factors p and q of ac whose sum or difference is equal to b.
3. Substitute the middle term bx by the sum or difference of px and qx.
4. Factorise the resulting expression by grouping.

The factorised form of a quadratic expression is
$$(mx + \mu)(nx + \lambda).$$
Not all quadratic expressions are factorable.

Completing the Square of a Quadratic Expression

$$ax^2 + bx + c = a\left\{x^2 + \frac{b}{a}x + \frac{c}{a}\right\}$$

Add and subtract $\left(\frac{b}{2a}\right)^2$ in the braces on the right hand side.

$$ax^2 + bx + c = a\left\{x^2 + \frac{b}{a}x + \left(\frac{b}{2a}\right)^2 - \left(\frac{b}{2a}\right)^2 + \frac{c}{a}\right\}$$

$$= a\left(x + \frac{b}{2a}\right)^2 + c - \frac{b^2}{4a}$$

$$\Rightarrow ax^2 + bx + c = a\left(x + \frac{b}{2a}\right)^2 + \frac{4ac - b^2}{4a}$$

Quadratic Equations

A quadratic equation is of the form $ax^2 + bx + c = 0$, where a, b and c are constants and $a \neq 0$.

Solutions to Quadratic Equations

(a) *Factorisation Method*

The factorised form of the quadratic equation $ax^2 + bx + c = 0$, is $(mx + \mu)(nx + \lambda) = 0$.

By the zero factor property,

Either $(mx + \mu) = 0 \Rightarrow x = -\frac{\mu}{m}$

or $(nx + \lambda) = 0 \Rightarrow x = -\frac{\lambda}{n}$.

(b) *Method of Completing the Square*

For quadratic equations whose factors are not eminent or which are not factorable, the method of completing the square can be used.

$$ax^2 + bx + c = 0$$
$$ax^2 + bx = -c$$

$$ax^2 + bx + \left(\frac{b}{2a}\right)^2 = \left(\frac{b}{2a}\right)^2 - \frac{c}{a}$$

$$\left(x + \frac{b}{2a}\right)^2 = \frac{b^2 - 4ac}{4a^2}$$

$$x + \frac{b}{2a} = \frac{\pm\sqrt{b^2 - 4ac}}{2a}$$

$$\Rightarrow x = \frac{-b \pm \sqrt{b^2 - 4ac}}{2a}, a \neq 0$$

(c) The Quadratic Formula Method

$$ax^2 + bx + c = 0 \Leftrightarrow x = \frac{-b \pm \sqrt{b^2 - 4ac}}{2a}, a \neq 0$$

(d) The Graphical Method

Provided the graph of the quadratic function
$y = ax^2 + bx + c$, crosses the x-axis, the intercepts
with the x-axis gives the solution of the equation
$ax^2 + bx + c = 0$.
The graph of the quadratic function
$y = ax^2 + bx + c$ is a parabola which has either a
maximum (a hill shape ∩) when $a < 0$ or a minimum (a
valley shape ∪) when $a > 0$.

Example

Find the solution of the following equations using the
graphical method.
(a) $x^2 + 2x - 3 = 0$ (b) $2 - x - x^2 = 0$

Solution

$y = x^2 + 2x - 3$

x	-4	-3	-2	-1	0	1	2
y	5	0	-3	-4	-3	0	5

$y = 2 - x - x^2$

x	-3	-2	-1	0	1	2
y	-4	0	2	2	0	-4

Solution

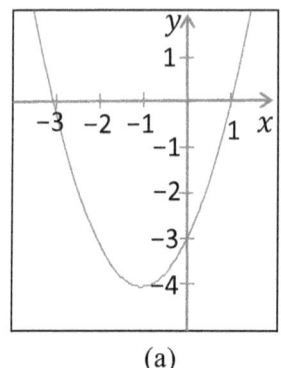
(a)

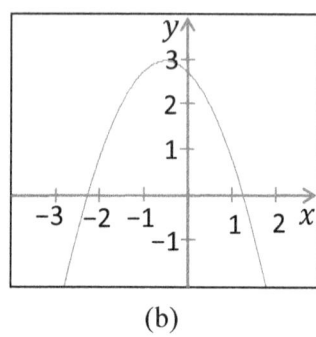
(b)

From graph (a) $x = -3$ or $x = 1$.
From graph (b) $x = -2.3$ or $x = 1.3$.

Nature of Roots of Quadratic Equations

The expression $b^2 - 4ac$ which appears under the square root sign in the quadratic formula is called the discriminant denoted by Δ. Thus $\Delta = b^2 - 4ac$. Clearly,

1. If $\Delta > 0$, the equation $ax^2 + bx + c = 0$, will have real and distinct roots
$$\alpha = \frac{-b-\sqrt{b^2-4ac}}{2a} \text{ and } \beta = \frac{-b+\sqrt{b^2-4ac}}{2a}.$$

2. If $\Delta = 0$, the equation $ax^2 + bx + c = 0$, will have real and equal (coincident) roots $\alpha = \beta = -\frac{b}{2a}$.

3. If $\Delta < 0$, the equation $ax^2 + bx + c = 0$, will have no real roots (roots will be imaginary or complex).

4. If $\Delta \geq 0$, the equation $ax^2 + bx + c = 0$, will have real roots.

Depending on the signs of Δ and a, the graph of the quadratic function $y = ax^2 + bx + c$ usually takes one of the following forms.

Δ and nature of roots	$a > 0$ minimum value	$a < 0$ maximum value
Real and distinct roots $\Delta > 0$		
Real and equal roots $\Delta = 0$		
Roots are not real (Imaginary or Complex roots) $\Delta < 0$		

Applications of Nature of Roots of Quadratic Equations

Let $y = ax^2 + bx + c$ be a quadratic function and $y = mx + k$ be the equation of a straight line. We can substitute $y = mx + k$ into $y = ax^2 + bx + c$ and rearrange to obtain $ax^2 + (b - m)x + (c - k) = 0$.

1. If $(b - m)^2 - 4a(c - k) > 0$, the line $y = mx + k$ intersects the curve $y = ax^2 + bx + c$ at two points.
2. If $(b - m)^2 - 4a(c - k) < 0$, the line $y = mx + k$ does not intersect the curve $y = ax^2 + bx + c$.
3. If $(b - m)^2 - 4a(c - k) = 0$, the line $y = mx + k$ is a tangent and so intersects the curve $y = ax^2 + bx + c$ at exactly one point.

Example
Find the values of m for which $y = mx - 3$, is a tangent to the curve $y = x^2 + 1$.

Solution
If $y = mx - 3$ is a tangent to the curve $y = x^2 + 1$, then the equation $x^2 + 1 = mx - 3$ has equal roots.
$\Rightarrow x^2 - mx + 4 = 0 \Rightarrow m^2 - 16 = 0 \Rightarrow m = \pm 4$

Symmetric Properties of Quadratic Functions

1. From $ax^2 + bx + c = a\left(x + \dfrac{b}{2a}\right)^2 + \dfrac{4ac - b^2}{4a}$, which we obtained under the section completing the square of a quadratic expression, $\left(x + \dfrac{b}{2a}\right)^2$ is a squared quantity. Hence, if $a > 0$, the quadratic function $y = ax^2 + bx + c$ will have a minimum value of $\dfrac{4ac - b^2}{4a}$ when $x = -\dfrac{b}{2a}$ and if $a < 0$, the function will have a maximum value of $\dfrac{4ac - b^2}{4a}$ when $x = -\dfrac{b}{2a}$.

2. The graph of $y = ax^2 + bx + c$ is symmetrical about the line $x = -\dfrac{b}{2a}$ called the axis of symmetry.

3. The roots $\alpha = \dfrac{-b - \sqrt{b^2 - 4ac}}{2a}$ and $\beta = \dfrac{-b + \sqrt{b^2 - 4ac}}{2a}$ are symmetrical about the line $x = -\dfrac{b}{2a}$.

4. The distance of each of the roots α and β from the axis of symmetry $x = -\dfrac{b}{2a}$ is $d = \dfrac{\sqrt{b^2 - 4ac}}{2a}$.

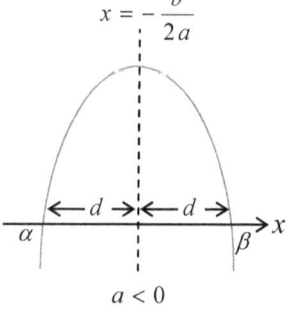

$a < 0$

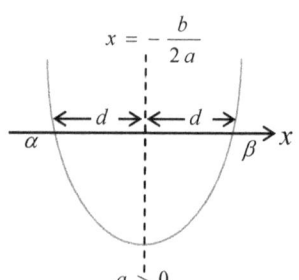

$a > 0$

5. The sum of roots, $s = \alpha + \beta = -\dfrac{b}{a}$ and the product of roots $p = \alpha\beta = \dfrac{c}{a}$.

Therefore, if the roots of the quadratic equation $ax^2 + bx + c = 0$ are known, or can be determined to be α and β, it is possible to find any other quadratic equation whose roots are symmetric functions of α and β such as α^2 and β^2, $\dfrac{1}{\alpha}$ and $\dfrac{1}{\beta}$, $\dfrac{1}{\alpha^2}$ and $\dfrac{1}{\beta^2}$, α^3 and β^3, $\dfrac{1}{\alpha^3}$ and $\dfrac{1}{\beta^3}$, etc.

Generally, we can write any quadratic equation in the form
$x^2 - (\text{sum of roots})x + \text{product of roots}$.

Example

Given that the roots of the equation $3x^2 + 8x - 3 = 0$ are α and β, form a quadratic equation with integral coefficients whose roots are

(i) α^2 and β^2 (ii) $\dfrac{\alpha}{\beta} + 1$ and $\dfrac{\beta}{\alpha} + 1$.

Solution

(i) $\alpha + \beta = -\dfrac{8}{3}$ and $\alpha\beta = -1$

$\alpha^2 + \beta^2 = (\alpha + \beta)^2 - 2\alpha\beta = \left(-\dfrac{8}{3}\right)^2 - 2(-1) = \dfrac{82}{9}$.

$\alpha^2\beta^2 = (\alpha\beta)^2 = (-1)^2 = 1$

Therefore, required equation is

$x^2 - \dfrac{82}{9}x + 1 = 0$ or $9x^2 - 82x + 9 = 0$

(ii) $\dfrac{\alpha}{\beta} + 1 + \dfrac{\beta}{\alpha} + 1 = \dfrac{\alpha^2 + \beta^2}{\alpha\beta} + 2 = \dfrac{82}{9} \div -1 + 2 = -\dfrac{64}{9}$

$\left(\dfrac{\alpha}{\beta} + 1\right)\left(\dfrac{\beta}{\alpha} + 1\right) = 2 + \dfrac{\alpha^2 + \beta^2}{\alpha\beta} = -\dfrac{64}{9}$

Therefore, required equation is

$x^2 + \dfrac{64}{9}x - \dfrac{64}{9} = 0$ or $9x^2 + 64x - 64 = 0$

The following identities are very useful especially when dealing with symmetric roots of equations which are functions of α^3 and β^3, $\dfrac{1}{\alpha^3}$ and $\dfrac{1}{\beta^3}$.

$$a^3 + b^3 = (a + b)\left(a^2 - 2ab + b^2\right)$$

$$a^3 - b^3 = (a - b)\left(a^2 + 2ab + b^2\right)$$

Transformations of Graphs of Functions

If k is a positive constant, then:
(i) The graph of $y = f(x) + k$ is a translation of $y = f(x)$, k units upward.
(ii) The graph of $y = f(x) - k$ is a translation of $y = f(x)$, k units downward.

9

(iii) The graph of $y = f(x + k)$ is a translation of $y = f(x)$ k units to the left.

(iv) The graph of $y = f(x - k)$ is a translation of $y = f(x)$, k units to the right.

(v) The graph of $y = -f(x)$ is a reflection of $y = f(x)$ in the x-axis.

(vi) The graph of $y = f(-x)$ is a reflection of $y = f(x)$ in the y-axis.

(vii) The graph of $y = kf(x)$ is a one-way stretch of $y = f(x)$ by factor k parallel to Oy or the y-direction.

(viii) The graph of $y = f(kx)$ is a one-way stretch of $y = f(x)$ $\dfrac{1}{k}$ parallel to Ox the x-direction.

Study the following transformations of the curve $f(x) = x^2 - 1$ noting value k in each case. This will help you appreciate better the transformations of graphs of functions.

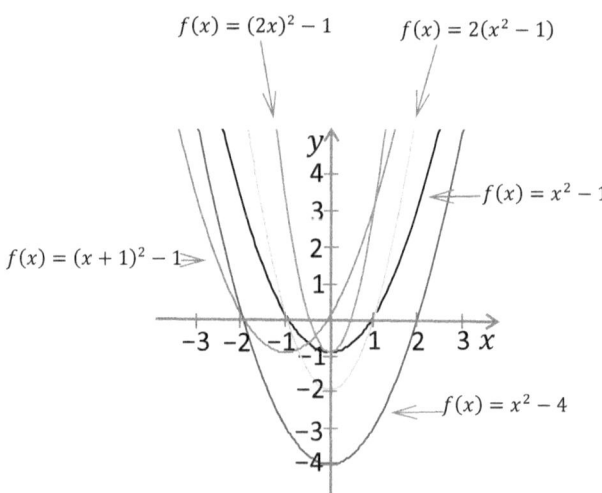

$f(x) = (2x)^2 - 1$ $\qquad$ $f(x) = 2(x^2 - 1)$

$f(x) = (x + 1)^2 - 1$

$f(x) = x^2 - 1$

$f(x) = x^2 - 4$

TOPIC 2
INEQUALITIES AND INEQUATIONS

An inequality is a statement which shows that two real quantities are not equal. An inequality which involves an unknown is called an inequation.

The Square of a Real Number
$(a + b)^2 \geq 0, (a - b)^2 \geq 0, \forall a, b \in \mathbb{R}$

Example
Show that $2x^2 + 3x + 5 > 0$.

Solution
By completing the square,

$$2x^2 + 3x + 5 = 2\left(x + \frac{3}{4}\right)^2 - \left(\frac{3}{4}\right)^2 + 5$$

$$= 2\left(x + \frac{3}{4}\right)^2 + \frac{71}{16}$$

$$\left(x + \frac{3}{4}\right)^2 \geq 0 \implies 2x^2 + 3x + 5 > 0$$

Solving Inequations

The method in which inequations are solved is the same as that in which equations are solved but for the fact that when the inequation is multiplied or divided by a negative number, the inequality sign changes sense from $<$ to $>$ or $\leq$ to $\geq$ and vice versa. Thus, if $a > b$, then

For all values of k, $a + k > b + k$ and $a - k > b - k$,

For all positive values of k, $ak > bk$ and $\frac{a}{k} > \frac{b}{k}$.

BUT for all negative values of k, $ak < bk$ and $\frac{a}{k} < \frac{b}{k}$.

11

Simple Linear Inequations

If a and b are positive real numbers then,

(i) $ax + b > 0 \Leftrightarrow x > -\dfrac{b}{a}.$

$-\dfrac{b}{a}$

(ii) $ax + b < 0 \Leftrightarrow x < -\dfrac{b}{a}.$

$-\dfrac{b}{a}$

(iii) $ax + b \geq 0 \Leftrightarrow x \geq -\dfrac{b}{a}.$

$-\dfrac{b}{a}$

(iv) $ax + b \leq 0 \Leftrightarrow x \leq -\dfrac{b}{a}.$

$-\dfrac{b}{a}$

Example
Solve the following and represent your answer on a number line. (a) $2x - 1 > 3$ (b) $3 - 2x \leq 5.$

Solution
(a) $2x - 1 > 3 \Longrightarrow x > 2$

2

(b) $3 - 2x \leq 5 \Longrightarrow x \geq -1$

−1

Quadratic Inequations

If the roots α and β of the quadratic equation $ax^2 + bx + c = 0$ are real and $\alpha < \beta$, then

(i) $ax^2 + bx + c > 0 \Leftrightarrow x < \alpha$ and $x > \beta$.

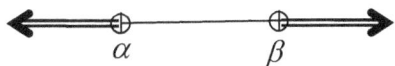

(ii) $ax^2 + bx + c < 0 \Leftrightarrow \alpha < x < \beta$.

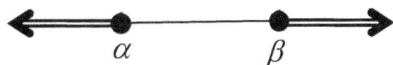

(iii) $ax^2 + bx + c \geq 0 \Leftrightarrow x \leq \alpha$ and $x \geq \beta$.

(iv) $ax^2 + bx + c \leq 0 \Leftrightarrow \alpha \leq x \leq \beta$.

Example

Solve the following inequalities and represent your answers on a number line.

(a) $x^2 - x - 6 > 0$ (b) $(3x - 2)(x + 5) \leq 0$.

Solution

(a) $x^2 - x - 6 > 0 \Rightarrow (x - 3)(x + 2) > 0$
$$x < -2 \; and \; x > 3$$

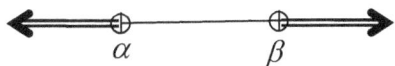

(b) $(3x - 2)(x + 5) \leq 0 \Rightarrow -5 \leq x \leq \frac{2}{3}$

13

Applications of Nature of Roots of Quadratic Equations

1. Find the range of values of k for which $2x^2 - kx + k = 0$ has real roots.

2. Find the range of values of the function $y = \frac{x^2}{x+1}$, $x \neq -1, x \in \mathbb{R}$.

3. Find the range of values of m for which the line $y = mx + 1$ intersects the curve $y = \frac{x}{x+1}$.

Solution

1. For real roots $b^2 - 4ac \geq 0$.
    $$\Rightarrow (-k)^2 - 4(2)(k) \geq 0 \Rightarrow k^2 - 8k \geq 0$$
    $$\Rightarrow k(k-8) \geq 0 \Rightarrow k \leq 0 \text{ and } k \geq 8$$

2. $y = \frac{x^2}{x+1} \Rightarrow x^2 - yx + y = 0,$
    $\forall x, x \neq -1, x \in \mathbb{R}, \ y^2 - 4y \geq 0 \Rightarrow y \leq 0 \text{ and } y \geq 4$

3. For intersection, $mx + 1 = \frac{x}{x+1} \Rightarrow mx^2 + mx + 1 = 0.$
    $\forall x \in \mathbb{R}, \ m^2 - 4m \geq 0 \Rightarrow m \leq 0 \text{ and } m \geq 4$

Inequations Involving Rational Functions

The critical values of $\frac{ax+b}{cx+d} > 0$ are $x_1 = -\frac{b}{a}$ and $x_2 = -\frac{c}{d}$.
To find the range of values of x which satisfy the inequality, determine which of the regions within the boundaries of the critical values satisfy the inequality.

Example

Find the solution of the inequality $\frac{3x+4}{2x-1} > 7.$

Solution

$$\frac{3x+4}{2x-1} - 7 > 0 \Rightarrow \frac{x-1}{2x-1} < 0.$$

Let $f(x) = \frac{x-1}{2x-1}$. The critical values are $\frac{1}{2}$ and 1.

14

	$x < \frac{1}{2}$	$\frac{1}{2} < x < 1$	$x > 1$
$(x - 1)$	$-$	$+$	$+$
$(2x - 1)$	$-$	$-$	$+$
$f(x)$	$+$	$-$	$+$

$$\therefore \frac{1}{2} < x < 1.$$

Compound Inequalities

Compound inequalities are inequalities of the form $f(x) < g(x) \leq h(x)$. They are best solved by separating them into the two components.

Example
Find the solution of the following inequlities.

(i) $3x + 4 < x^2 - 6 \leq 9 - 2x$

(ii) $(x + 1)(x - 3)(2x - 5) > 0$

Solution
(i)   The two sections are solved independently.

$3x + 4 < x^2 - 6$	$x^2 - 6 \leq 9 - 2x$
$x^2 - 3x - 10 > 0$	$x^2 + 2x - 15 \leq 0$
$(x - 5)(x + 2) > 0$	$(x + 5)(x - 3) \leq 0$
$x < -2$ and $x > 5$	$-5 \leq x \leq 3$

(ii) $(x + 1)(x - 3)(2x - 5) > 0$

Let $f(x) = (x + 1)(x - 3)(2x - 5)$

The critical values are $x = -1, x = 3, x = \frac{5}{2}$

	$x < -1$	$-1 < x < \frac{5}{2}$	$\frac{5}{2} < x < 3$	$x > 3$
$(x + 1)$	$-$	$+$	$+$	$+$
$(2x - 5)$	$-$	$-$	$+$	$+$
$(x - 3)$	$-$	$-$	$-$	$+$
$f(x)$	$-$	$+$	$-$	$+$

$$\therefore -1 < x < \frac{5}{2} \text{ and } x > 3$$

15

Modulus or Absolute Value Inequalities

A modulus or absolute value inequality is of the form
$f(x) = |x|$ and is defined for all $x \in \mathbb{R}$ as,

$$f(x) = |x| = \sqrt{x^2} \Rightarrow f(x) = \begin{cases} x, & x > 0 \\ 0, & x = 0 \\ -x, & x < 0 \end{cases}$$

Absolute Value Properties

(i) $|x| = a \Rightarrow x = -a \text{ or } x = a.$

(ii) $|x| \leq a \Rightarrow -a \leq x \leq a.$

(iii) $|x| > a \Rightarrow x < -a \text{ or } x > a.$

These properties are used to solve absolute value inequations.

Example
Solve the following inequalities.
  (i) $3|x - 1| < |x - 3|$ (ii) $x + 6 > |2x + 3|$

(iii) $x^2 - |x| - 6 < 0$ (iv) $\left|\frac{x+3}{x-1}\right| \geq \frac{x+3}{x-1}$

Solution
(a) (i) $3|x - 1| < |x - 3| \Rightarrow 9(x^2 - 2x + 1) < x^2 - 6x + 9$

$\Rightarrow 2x^2 - 3x < 0 \Rightarrow x(2x - 3) < 0 \Rightarrow 0 < x < \dfrac{3}{2}$

(ii) $x + 6 > |2x + 3| \Rightarrow x^2 + 12x + 36 > 4x^2 + 12x + 9$

$\Rightarrow x^2 < 9 \Rightarrow -3 < x < 3$

(iii) $x^2 - |x| - 6 < 0$

$x > 0 \Rightarrow x^2 - x - 6 < 0 \Rightarrow (x - 3)(x + 2) < 0$

$\Rightarrow -2 < x < 3.$

$x < 0 \Rightarrow x^2 + x - 6 < 0 \Rightarrow (x + 3)(x - 2) < 0$

$\Rightarrow -3 < x < 2.$

Combining the solutions we have $-3 < x < 3$.

(iv) $\left|\dfrac{x+3}{x-1}\right| \geq \dfrac{x+3}{x-1}$

Suppose $\dfrac{x+3}{x-1} > 0$, then $\dfrac{x+3}{x-1} \geq \dfrac{x+3}{x-1}$ which is not real.

Suppose $\dfrac{x+3}{x-1} < 0$, then $-\dfrac{x+3}{x-1} > \dfrac{x+3}{x-1} \implies \dfrac{x+3}{x-1} < 0$.

Critical values are $x = -3$ and $x = 1$.

	$x \leq -3$	$-3 \leq x \leq 1$	$x \geq 1$
$(x+3)$	$-$	$+$	$+$
$(x-1)$	$-$	$-$	$+$
$f(x)$	$+$	$-$	$+$

$\therefore\ -3 \leq x \leq 1$

Graphs of Absolute Value Functions

1. **The Graph of $y = a|x - h| + k$**

   When $x = h, y = k$. There are two cases to examine.

   (i) If $a > 0$ the graph of $y = a|x - h| + k$ is a symmetrical "V" with minimum point (h, k), gradient $m = a$ on the right side of minimum point for which $x > h$ and gradient $m = -a$ on the left side of minimum point for which $x < h$.

   (ii) If $a < 0$ the graph of $y = a|x - h| + k$ is an upside down symmetrical "Λ" with maximum point (h, k), gradient $m = -a$ when $x > h$ and $m = a$ when $x < h$.

To sketch the graph of $y = a|x - h| + k$, sketch the graph of $y = a(x - h) + k$ and reflect it from the point (h, k).

Example

Sketch the following graphs.

(a) $y = 2|x + 3| + 1$ (b) $y = -2|x + 3| + 1$

Solution

(a) $a = 2 > 0 \implies y = 2|x + 3| + 1$ has a minimum point $(-3, 1)$ and its gradient is 2 when $x > -3$ and -2 when $x < -3$.

(b) $a = -2 < 0 \Rightarrow y = 2|x + 3| + 1$ has a maximum point $(-3,1)$ and its gradient is -2 when $x > -3$ and 2 when $x < -3$.

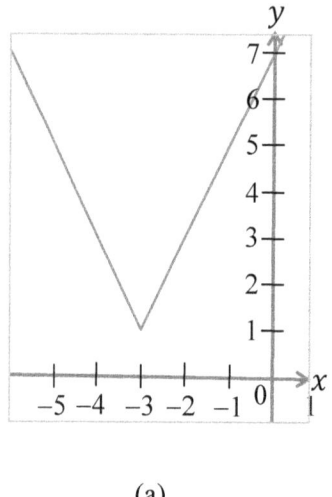

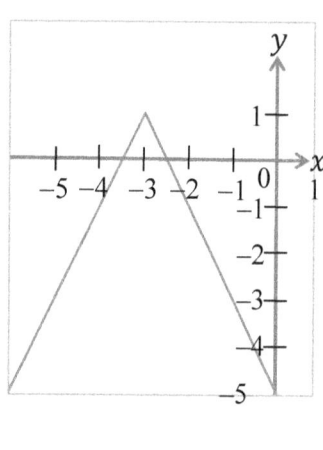

(a)   (b)

2. **The Graph of $y = a|(x - \alpha)(x - \beta)| + k$**

When $x = \alpha$ or $x = \beta$, $y = k$. There are two cases to examine.

(i) If $a > 0$, the graph of $y = a|(x - \alpha)(x - \beta)| + k$ is a symmetrical "W" with a hill shape parabola whose maximum point is between the minimum points (α, k) and (β, k).

(ii) If $a < 0$, the graph of $y = a|(x - \alpha)(x - \beta)| + k$ is a symmetrical "M" with a valley shape parabola between the maximum points (α, k) and (β, k).

To sketch the graph of $y = a|(x - \alpha)(x - \beta)| + k$, sketch the parabola $y = a(x - \alpha)(x - \beta) + k$ and reflect the section between the minimum or maximum points (α, k) and (β, k).

Example

Sketch the following graphs.

(a) $y = 2|(x - 2)(x + 1)| - 4$

(b) $y = -2|(x - 2)(x + 1)| - 4$

18

Solution

(a) $a = 2 > 0 \implies y = 2|(x-2)(x+1)| - 4$ is a symmetrical "W" with a hill shape parabola whose maximum point is between its minimum points $(-1,-4)$ and $(2,-4)$.

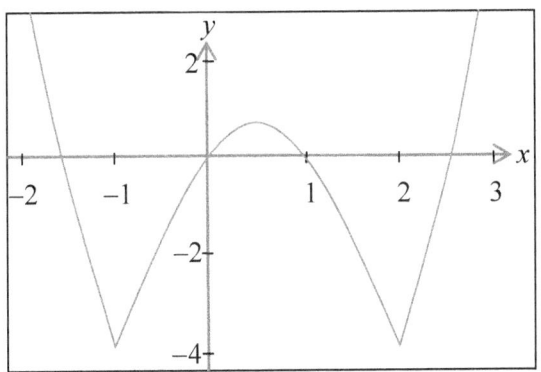

(b) $a = -2 < 0 \implies y = -2|(x-2)(x+1)| - 4$ is a symmetrical "M" with a hill shape parabola whose minimum point is between its maximum points $(-1,-4)$ and $(2,-4)$.

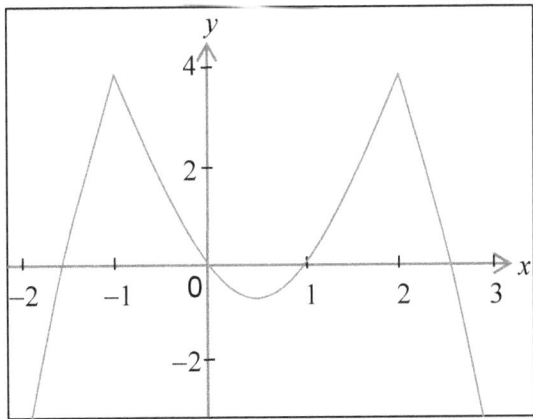

Solving Modulus Inequalities Using the Graphical Method

Example

Solve the following inequalities using the graphical method.

(i) $3|x - 1| < |x - 3|$ (ii) $x + 6 > |2x + 3|$

(iii) $x^2 - |x| - 6 < 0$ (iv) $\left|\frac{x+3}{x-1}\right| \geq \frac{x+3}{x-1}$

(i)   From the sketch below the graph of $y = |x - 3|$ lies

above the graph of $y = 3|x - 1|$ between 0 and $\frac{3}{2}$.

Therefore, $0 < x < \frac{3}{2}$.

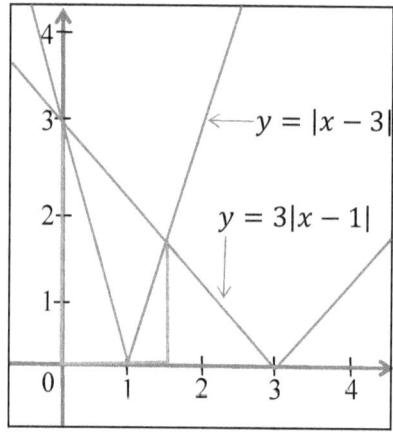

(ii) From the sketch below the graph of $y = x + 6$ lies
above the graph of $y = |2x + 3|$ between -3 and 3.
Therefore, $-3 < x < 3$.

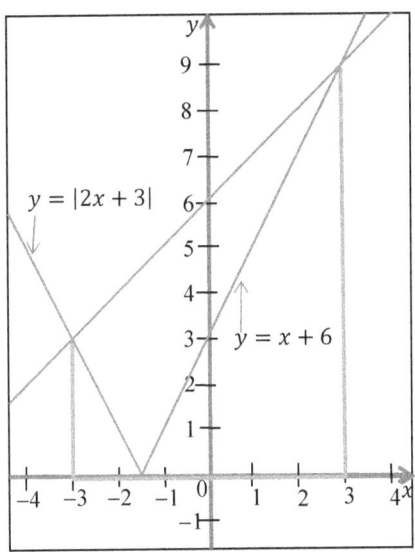

(iii) From the sketch below the graph of $y = x^2 - |x| - 6$ lies below the x-axis between -3 and 3. Therefore, $-3 < x < 3$.

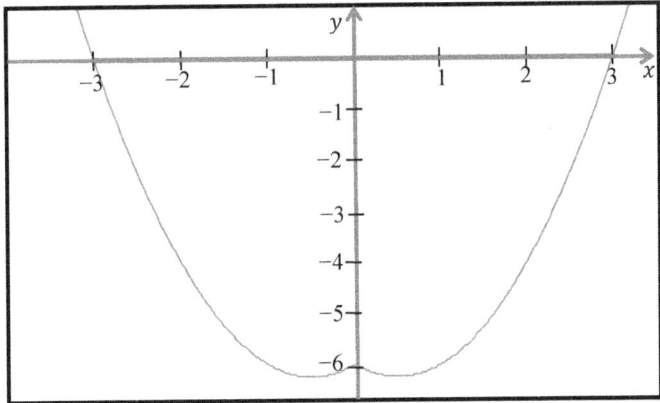

(iv) The graph of $y = \left|\dfrac{x+3}{x-1}\right|$ lies above the graph of $y = \dfrac{x+3}{x-1}$

between -3 and 1. Therefore, $-3 \leq x \leq 1$.

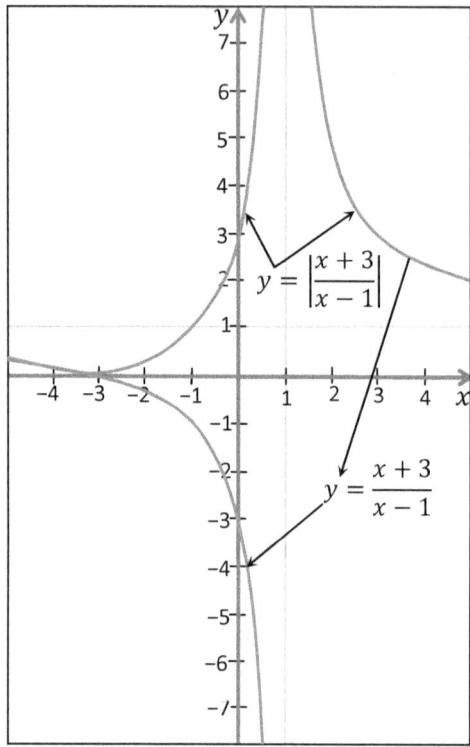

TOPIC 3
INDICES, LOGARITHMS AND SURDS

INDICES

$$a^n = a \times a \times a \times a.............. \times a \text{ up to } n \text{ times}$$

Laws of Indices

1.	**Exponent 1**	$a^1 = a, \ a \neq 0$
2.	**Exponent 0**	$a^0 = 1, \ a \neq 0$
3.	**Product**	$a^m \times a^n = a^{m+n}$
4.	**Quotient**	$a^m \div a^n = a^{m-n}$
5.	**Negative Exponent**	$a^{-n} = \dfrac{1}{a^n}$
6.	**Power**	$\left(a^m\right)^n = a^{mn}$
7.	**Power of a Product**	$(ab)^n = a^n b^n$
8.	**Power of a Quotient**	$\left(\dfrac{a}{b}\right)^n = \dfrac{a^n}{b^n}$
9.	**Fractional Exponents** (m and n are integers and $n > 0, a > 0$)	$a^{\frac{1}{n}} = \sqrt[n]{a}$ $a^{\frac{m}{n}} = \left(a^{\frac{1}{n}}\right)^m = \left(\sqrt[n]{a}\right)^m$

Note!!

If n is even and $a < 0$, $\sqrt[n]{a}$ does not exist.

These laws can be used to solve exponential equations and simplify exponential expressions.

Example

Solve the equation $2^{2x} - 3(2^x) - 4 = 0$

23

Solution

$$(2^x)^2 - 3(2^x) - 4 = 0 \Rightarrow (2^x - 4)(2^x + 1) = 0$$
$$2^x = 2^2 \Rightarrow x = 2 \text{ or } 2^x = -1 \text{ (not real)}$$

LOGARITHMS

Definition: The logarithm of a number n to the base b is the power p to which b must be raised to give n.

i.e. $\log_b n = p \Leftrightarrow n = b^p$, $n > 0$

By this definition, $b^{\log_b n} = n$, $\log_b b = 1$ and $\log_b 1 = 0$.

Laws of Logarithms

(i) $\log_b xy = \log_b x + \log_b y$

(ii) $\log_b\left(\dfrac{x}{y}\right) = \log_b x - \log_b y$

(iii) $\log_b x^n = n\log_b x$

(iv) **Change of base Formula:** $\log_b n = \dfrac{\log_a n}{\log_a b}$

We can use these laws to solve logarithmic equations and simplify logarithmic expressions.

Example
Solve the equation:
(a) $\log_4 8 + \log_4(x + 5) = 3$
(b) $\log_5(x + 3) - \log_5(x - 7) = 3$
(c) $3\log_7 x = 2\log_x 7 + 5$
(d) $4^{2x} - 4^{x+2} = 80$
(e) $\log(x - 2) + \log 2 = 2\log y$
    $\log(x - 3y + 3) = 0$

Solution

(a) $\log_4 8 + \log_4(x + 5) = 3$

$\log_4 8(x + 5) = 3$

$8x + 40 = 64 \Rightarrow x = 3$

(b) $\log_5(x + 3) - \log_5(x - 7) = 3$

$\log_5 \left(\dfrac{x + 3}{x - 7}\right) = 3$

$\Rightarrow \dfrac{x + 3}{x - 7} = 125$

$x + 3 = 125x - 875$

$124x = 878 \Rightarrow x = \dfrac{439}{62}$

(c) $3 \log_7 x = 2 \log_x 7 + 5$

$\dfrac{3}{\log_x 7} = 2 \log_x 7 + 5$

$2(\log_x 7)^2 + 5 \log_x 7 - 3 = 0$

$(2 \log_x 7 - 1)(\log_x 7 + 3) = 0$

$\Rightarrow x = 49$ or $\log_x 7 = -3 \Rightarrow x = 7^{-\frac{1}{3}}$

(d) $4^{2x} - 4^{x+2} = 80$

$(4^x)^2 - 16(4^x) - 80 = 0$

$(4^x - 20)(4^x + 4) = 0$

$4^x = 20 \Rightarrow x = \log_4 20$ or $x = \log_4(-4)$ (not real)

(e) $\log(x - 2) + \log 2 = 2 \log y \Rightarrow y^2 = 4x - 4 \ldots \ldots \textcircled{1}$

$\log(x - 3y + 3) = 0 \Rightarrow x = 3y - 2 \ldots \ldots \ldots \ldots \ldots \textcircled{2}$

Substitute $\textcircled{2}$ in $\textcircled{1}$: $y^2 - 6y + 8 = 0$

$(y - 2)(y - 4) = 0 \Rightarrow y = 2$ or $y = 4$

$y = 2 \Rightarrow x = 4$ or $y = 4 \Rightarrow x = 10$

Natural Logarithms

A natural logarithm is a logarithm whose base is e, where

$e = 2.71828..$ to five decimal places.

i.e. $\log_e n = p \Leftrightarrow n = e^p$

We usually write $\log_e x$ simply as $\ln x$, read "elen x".
By this definition, $a^x = e^{x \ln a}$.
All the laws for common logarithms hold for natural logarithms, and can be used to solve natural logarithmic equations and simplify natural logarithmic expressions.

Example

Solve the equation:

    (a) $e^{2 \ln x} + x^2 = 8$

    (b) $2e^x + 5 = 3e^{-x}$

    (c) $(\ln x)^2 - \ln x - 6 = 0$

Solution

    (a) $e^{2 \ln x} + x^2 = 8 \Longrightarrow 2x^2 = 8 \Longrightarrow x = \pm 2$

    (b) $2e^x + 5 = 3e^{-x} \Longrightarrow 2e^{2x} + 5e^x - 3 = 0$

        $(2e^x - 1)(e^x + 3) = 0$

        $e^x = \dfrac{1}{2}$ or $e^x = -3$

        $x = -\ln 2$ or $x = \ln -3$ (Not real)

    (c) $(\ln x)^2 - \ln x - 6 = 0 \Longrightarrow (\ln x - 3)(\ln x + 2) = 0$

        $\rightarrow x = e^3$ or $x = e^{-2}$

Graphs of Exponential and Logarithmic Functions

To plot the graphs of the exponential functions $y = b^x$ (e.g. $y = 2^x$) and the logarithmic function $y = \log_b x$ (e.g. $y = \log_{10} x$), make a table of values of y against x.

The following graph shows the typical characteristics of these types of graphs.

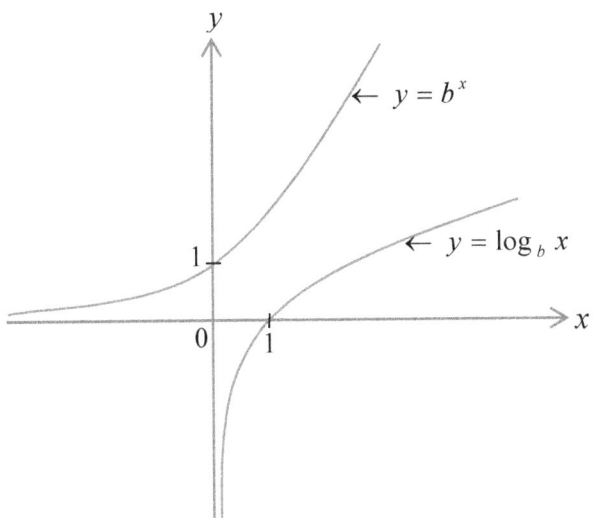

SURDS

1. $\sqrt[n]{ab} = \left(\sqrt[n]{a}\right)\left(\sqrt[n]{b}\right)$ 2. $\sqrt[n]{\dfrac{a}{b}} = \dfrac{\sqrt[n]{a}}{\sqrt[n]{b}}$

3. $\sqrt[n]{a^m} = \left(\sqrt[n]{a}\right)^m = a^{\frac{m}{n}}$

The conjugate of $a + \sqrt{b}$ is $a - \sqrt{b}$.

$$\left(a + \sqrt{b}\right)\left(a - \sqrt{b}\right) = a^2 - b$$

Rationalising the denominator is the process of removing the surd from the denominator of a fractional surd expression by multiplying both the denominator and numerator by the conjugate of the denominator.

e.g. $\dfrac{1}{2\sqrt{3}-1} = \left(\dfrac{1}{2\sqrt{3}-1}\right)\left(\dfrac{2\sqrt{3}+1}{2\sqrt{3}+1}\right) = \dfrac{2\sqrt{3}+1}{\left(2\sqrt{3}\right)^2 - 1^2} = \dfrac{2\sqrt{3}+1}{11}$

NOTE!! $\sqrt[n]{a} \pm \sqrt[n]{b} \neq \sqrt[n]{a \pm b}$

TOPIC 4
POLYNOMIALS WITH REAL COEFFICIENTS

Definition

A polynomial expression of degree n is an expression of the form $a_n x^n + a_{n-1} x^{n-1} + a_{n-2} x^{n-2} + \ldots + a_1 x^1 + a_0$, where $a_n \neq 0$ and $n \in \mathbb{Z}^+$.

Classification Common Polynomials

Value of n	Polynomial Name	Example
1	Linear	$3x + 1$
2	Quadratic	$2x^2 + 5x - 3$
3	Cubic	$x^3 - 2x^2 - 5x + 14$
4	Quartic	$2x^4 + 6x^3 - x^2 + 5x - 4$
5	Quintic	$x^5 + 2x^4 - 3x^3 - x^2 + 3x - 1$

A polynomial with one term is called a monomial.
A polynomial with two terms is called a binomial.
A polynomial with three terms is called a trinomial.

Algebra of Polynomial Functions

A polynomial function is of the form

$$y = a_n x^n + a_{n-1} x^{n-1} + a_{n-2} x^{n-2} + \ldots + a_1 x^1 + a_0$$

1. To add or subtract two or more polynomials, simply add or subtract like terms.
2. To multiply polynomials, use the distributive law to multiply each term of one polynomial by the terms of the other polynomial.
3. To divide two polynomials use long division (the division algorithm).

The Remainder Theorem

The remainder theorem states that when a polynomial $f(x)$ is divided by $ax+b$, the remainder is $f\left(-\dfrac{b}{a}\right)$.

Note!!

The remainder is always of lesser degree than the divisor.

The Factor Theorem

The factor theorem states that if the remainder when $f(x)$ is divided by $ax+b$ is zero, then $ax+b$ is a factor of $f(x)$.

Conversely, if $f\left(-\dfrac{b}{a}\right) = 0$ then $ax+b$ is a factor of the polynomial $f(x)$.

Applications of the Remainder and Factor Theorems

We can use the remainder and factor theorems to factorise polynomials or to solve polynomial equations. Hence given a polynomial $p(x)$ of degree n,

1. $P(x) = (x-a)\,Q(x) + R$, $Q(x)$ is of degree $n-1$.
2. $P(x) = (x-a)(x-b)\,Q(x) + Ax + B$, $Q(x)$ is of degree $n-2$.
   $\Rightarrow P(a) = Aa + B$ and $P(b) = Ab + B$

Example

Given that $(x+3)$ is a factor of $f(x) = x^3 + 6x^2 + kx + 6$. Find the value of k, hence, factorise the expression completely.

Solution

$(x+3)$ is a factor $\Rightarrow f(-3) = 0$

$\Rightarrow (-3)^3 + 6(-3)^2 - 3k + 6 = 0 \Rightarrow k = 11$.

$\Rightarrow f(x) = (x+3)(x^2 + px + 2) = x^3 + 6x^2 + 11x + 6.$

Equating coefficients, $3p + 2 = 11 \Rightarrow p = 3$.

$f(x) = (x+3)(x^2 + 3x + 2) \Rightarrow f(x) = (x+3)(x+2)(x+1).$

29

TOPIC 5
PARTIAL FRACTIONS

If the fraction is an improper fraction, first divide to express it as a whole polynomial and a proper fraction.

1. To any linear denominator factor $ax + b$ use the partial fraction $\dfrac{A}{ax + b}$.

2. To any repeated factor $(ax + b)^2$ in the denominator use the partial fractions $\dfrac{A}{ax + b} + \dfrac{B}{(ax + b)^2}$.

3. To any irreducible quadratic factor $ax^2 + bx + c$ in the denominator use the partial fraction $\dfrac{Ax + B}{ax^2 + bx + c}$.

The Cover-Up Method

The cover-up method works only for linear factors.

1. $f(x) = \dfrac{\cdots}{(ax + b)\cdots}$

   $\Rightarrow f(x) = \dfrac{f\left(-\dfrac{b}{a}\right) \text{ with } (ax + b) \text{ covered - up}}{ax + b} + \cdots$

2. $f(x) = \dfrac{\cdots}{(ax + b)^2 \cdots}$

   $\Rightarrow f(x) = \dfrac{f\left(-\dfrac{b}{a}\right) \text{ with } (ax + b)^2 \text{ covered - up}}{(ax + b)^2} + \dfrac{A}{(ax + b)} \cdots$

   We then find the value of A.

TOPIC 6
TRIGONOMETRY

The Pythagoras Theorem

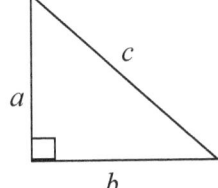

$$c^2 = a^2 + b^2 \Rightarrow c = \sqrt{a^2 + b^2}$$

$$a^2 = c^2 - b^2 \Rightarrow a = \sqrt{c^2 - b^2}$$

$$b^2 = c^2 - a^2 \Rightarrow b = \sqrt{c^2 - a^2}$$

The Six Trigonometric Ratios

Trigonometric Ratio	Reciprocal Trigonometric Ratio
$\sin\theta = \dfrac{\text{opp}}{\text{hyp}} = \dfrac{a}{c}$	$\csc\theta = \dfrac{\text{hyp}}{\text{opp}} = \dfrac{c}{a} = \dfrac{1}{\sin\theta}$
$\cos\theta = \dfrac{\text{adj}}{\text{hyp}} = \dfrac{b}{c}$	$\sec\theta = \dfrac{\text{hyp}}{\text{adj}} = \dfrac{c}{b} = \dfrac{1}{\cos\theta}$
$\tan\theta = \dfrac{\text{opp}}{\text{adj}} = \dfrac{a}{b}$	$\cot\theta - \dfrac{\text{adj}}{\text{opp}} - \dfrac{b}{a} = \dfrac{1}{\tan\theta}$

Special Angle Trigonometric Ratios

θ	0°	30°	45°	60°	90°
$\sin\theta$	0	$\dfrac{1}{2}$	$\dfrac{\sqrt{2}}{2}$	$\dfrac{\sqrt{3}}{2}$	1
$\cos\theta$	1	$\dfrac{\sqrt{3}}{2}$	$\dfrac{\sqrt{2}}{2}$	$\dfrac{1}{2}$	0
$\tan\theta$	0	$\dfrac{\sqrt{3}}{3}$	1	$\sqrt{3}$	∞

θ	0°	30°	45°	60°	90°
$\mathrm{cosec}\theta$	∞	2	$\sqrt{2}$	$\dfrac{2\sqrt{3}}{3}$	1
$\sec\theta$	1	$\dfrac{2\sqrt{3}}{3}$	$\sqrt{2}$	2	∞
$\cot\theta$	∞	$\sqrt{3}$	1	$\dfrac{\sqrt{3}}{3}$	0

Radian Measure

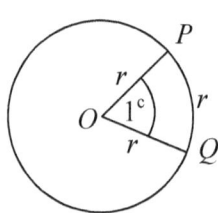

A radian is the angle subtended at the centre of a circle by an arc whose length is equal to the radius of the circle.

$\Rightarrow 360° = 2\pi$ or $180° = \pi$

We can use the above relation to convert degrees to radians and vice versa.

e.g. $135° = \dfrac{135°}{360°} \times 2\pi = \dfrac{3\pi}{4}$.

$\dfrac{7\pi}{5} = \dfrac{\left(\dfrac{7\pi}{5}\right)}{2\pi} \times 360 = 252°$

Arc Length, Area of a Sector and Area of a Segment

	θ in radians	θ in degrees
Arc length, l	$l = r\theta$	$l = \dfrac{\theta}{360} \times 2\pi r$
Area of Sector, S	$S = \dfrac{1}{2}r^2\theta$	$S = \dfrac{\theta}{360} \times \pi r^2$
Area of Segment, A	$S = \dfrac{1}{2}r^2(\theta - \sin\theta)$	$A = \dfrac{\theta}{360} \times \pi r^2 - \dfrac{1}{2}r^2\sin\theta$

The General Angle

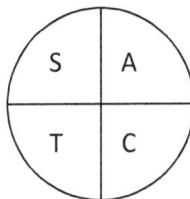

The figure on the left shows that **ALL** the trigonometric ratios are positive in the first quadrant and **S**ine, **T**angent and **C**osine are positive in the second, third and fourth quadrants respectively.

Conventionally, angles in the anti-clockwise direction are regarded as positive while angles in the clockwise direction are regarded as negative. $-\theta = 360 - \theta$.

$$\sin\theta = \sin(180 - \theta) = -\sin(-\theta) = -\sin(360 - \theta)$$
$$\cos\theta = -\cos(180 - \theta) = \cos(-\theta) = \cos(360 - \theta)$$
$$\tan\theta = -\tan(180 - \theta) = -\tan(-\theta) = -\tan(360 - \theta)$$

Simple Trigonometric Equations

$$\sin\theta = k \implies \theta = \sin^{-1} k$$

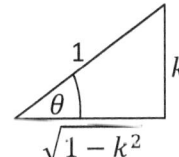

$$\implies \cos\theta = \sqrt{1 - k^2}$$

$$\text{and } \tan\theta = \frac{k\sqrt{1 - k^2}}{1 - k^2}$$

$$\cos\theta = k \implies \theta = \cos^{-1} k$$

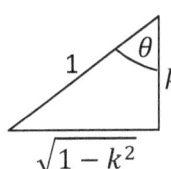

$$\implies \sin\theta = \sqrt{1 - k^2}$$

$$\text{and } \tan\theta = \frac{\sqrt{1 - k^2}}{k}$$

$$\tan\theta = k \implies \theta = \tan^{-1} k$$

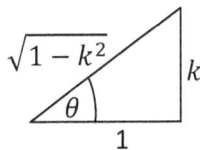

$$\implies \sin\theta = \frac{k\sqrt{1 + k^2}}{1 + k^2}$$

$$\text{and } \cos\theta = \frac{\sqrt{1 + k^2}}{1 + k^2}$$

General Solution of Trigonometric Equations

Equation	in degrees	in radians
$\sin\theta = \sin\alpha$	$\theta = 180n + (-1)^n \alpha$	$\theta = n\pi + (-1)^n \alpha$
$\cos\theta = \cos\alpha$	$\theta = 360n \pm \alpha$	$\theta = 2n\pi \pm \alpha$
$\tan\theta = \tan\alpha$	$\theta = 180n + \alpha$	$\theta = n\pi + \alpha$

α, is the associated acute angle.

Trigonometric Identities

Basic Identities

$$\tan\theta = \frac{\sin\theta}{\cos\theta} \qquad\qquad \cot\theta = \frac{\cos\theta}{\sin\theta}$$

Reciprocal Identities

$$\operatorname{cosec}\theta = \frac{1}{\sin\theta}, \qquad \sec\theta = \frac{1}{\cos\theta}, \qquad \cot\theta = \frac{1}{\tan\theta}$$

Complementary Angle Identities

$$\sin\theta = \cos(90 - \theta)° = \cos\left(\frac{\pi}{2} - \theta\right)$$

$$\cos\theta = \sin(90 - \theta)° = \sin\left(\frac{\pi}{2} - \theta\right)$$

$$\tan\theta = \cot(90 - \theta)° = \cot\left(\frac{\pi}{2} - \theta\right)$$

$$\cot\theta = \tan(90 - \theta)° = \tan\left(\frac{\pi}{2} - \theta\right)$$

Pythagorean Identities

$$\cos^2\theta + \sin^2\theta = 1$$
$$1 + \tan^2\theta = \sec^2\theta$$
$$1 + \cot^2\theta = \operatorname{cosec}^2\theta$$

Compound Angle Identities

$$\sin(A+B) = \sin A\cos B + \sin B\cos A$$
$$\sin(A-B) = \sin A\cos B - \sin B\cos A$$
$$\cos(A+B) = \cos A\cos B - \sin B\sin A$$
$$\cos(A-B) = \cos A\cos B + \sin B\sin A$$

$$\tan(A+B) = \frac{\tan A + \tan B}{1 - \tan A\tan B} \qquad \tan(A-B) = \frac{\tan A - \tan B}{1 + \tan A\tan B}$$

Double Angle Identities

$$\sin 2A = 2\sin A\cos A,$$

$$\cos 2A = \begin{cases} \cos^2 A - \sin^2 A \\ 1 - 2\sin^2 A \\ 2\cos^2 A - 1 \end{cases} \Rightarrow \begin{cases} \cos^2 A = \dfrac{1}{2}\left(1 + \cos 2A\right) \\ \sin^2 A = \dfrac{1}{2}\left(1 - \cos 2A\right) \end{cases}$$

$$\tan 2A = \frac{2\tan A}{1 - \tan^2 A}.$$

We can also write the above identities as follows.

$$\sin\theta = 2\sin\left(\frac{\theta}{2}\right)\cos\left(\frac{\theta}{2}\right),$$

$$\cos\theta = \begin{cases} \cos^2\left(\dfrac{\theta}{2}\right) - \sin^2\left(\dfrac{\theta}{2}\right) \\ 1 - 2\sin^2\left(\dfrac{\theta}{2}\right) \\ 2\cos^2\left(\dfrac{\theta}{2}\right) - 1 \end{cases} \Rightarrow \begin{cases} \cos^2\left(\dfrac{\theta}{2}\right) = \dfrac{1}{2}\left(1 + \cos\theta\right) \\ \sin^2\left(\dfrac{\theta}{2}\right) = \dfrac{1}{2}\left(1 - \cos\theta\right) \end{cases}$$

$$\tan \theta = \frac{2 \tan \left(\dfrac{\theta}{2} \right)}{1 - \tan^2 \left(\dfrac{\theta}{2} \right)} \quad \ldots\ldots\ldots \quad \ldots\ldots \quad * \, .$$

Half Angle Identities

Let $t = \tan \left(\frac{a}{2} \right)$ then from $*$, $\tan A = \frac{2t}{1-t^2}$.

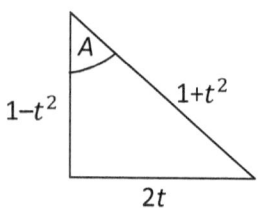

$$\Rightarrow \sin A = \frac{2t}{1+t^2}$$

$$\text{and } \cos A = \frac{1-t^2}{1+t^2} \, .$$

Factor Formulae

$$\sin A + \sin B = 2 \sin \left(\frac{A+B}{2} \right) \cos \left(\frac{A-B}{2} \right)$$

$$\sin A - \sin B = 2 \cos \left(\frac{A+B}{2} \right) \sin \left(\frac{A-B}{2} \right)$$

$$\cos A + \cos B = 2 \cos \left(\frac{A+B}{2} \right) \cos \left(\frac{A-B}{2} \right)$$

$$\cos A - \cos B = -2 \sin \left(\frac{A+B}{2} \right) \sin \left(\frac{A-B}{2} \right)$$

The Expression $a\cos\theta + b\sin\theta$

$a\cos\theta \pm b\sin\theta = R\cos(\theta \pm \alpha)$

where $R = \sqrt{a^2 + b^2}$ and $\alpha = \tan^{-1}\left(\dfrac{b}{a}\right)$

$a\cos\theta \pm b\sin\theta = R\sin(\theta \pm \alpha)$

where $R = \sqrt{a^2 + b^2}$ and $\alpha = \tan^{-1}\left(\dfrac{a}{b}\right)$

The Maximum and Minimum Values of $a\cos\theta + b\sin\theta$

Since $-1 \le \cos(\theta \pm \alpha) \le 1$ and $-1 \le \sin(\theta \pm \alpha) \le 1$

$-R \le a\cos\theta \pm b\sin\theta \le R$, where $R = \sqrt{a^2 + b^2}$

Example

Express $3\cos x + 4\sin x$ in the form $R\cos(x - \alpha)$, where R is a constant and α is acute. Hence, find the greatest and least values of $3\cos x + 4\sin x$.

Solution

$R\cos(x - \alpha) \equiv R\cos\alpha\cos x + R\sin\alpha\sin x$.

Comparing coefficients of RHS with $3\cos x + 4\sin x$, we have

$\left.\begin{array}{l} R\sin\alpha = 4 \\ R\cos\alpha = 3 \end{array}\right\} \Rightarrow \alpha = \tan^{-1}\left(\dfrac{4}{3}\right) = 53.1°, R = \sqrt{4^2 + 3^2} = 5$

$\Rightarrow 3\cos x + 4\sin x \equiv 5\cos(x - 53.1°)$.

$-1 \le \cos(x - 53.1°) \le 1 \Rightarrow -5 \le 5\cos(x - 53.1°) \le 5$

$\Rightarrow -5 \le 3\cos x + 4\sin x \le 5$.

Therefore, the minimum value of $3\cos x + 4\sin x$ is -5 and the maximum value is 5.

Graphs of Trigonometric Functions

$y = \sin \theta$

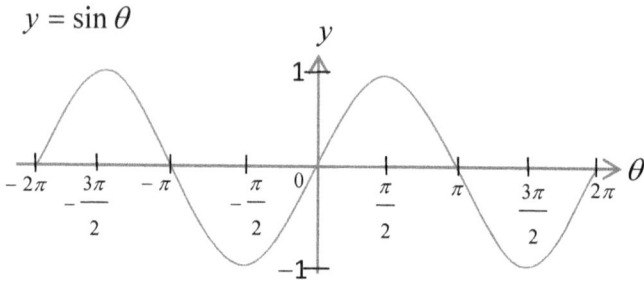

$y = \cos \theta$

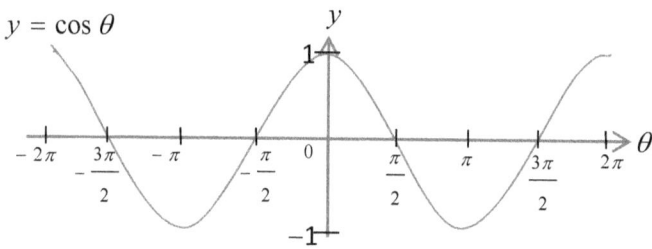

$y = \tan \theta$

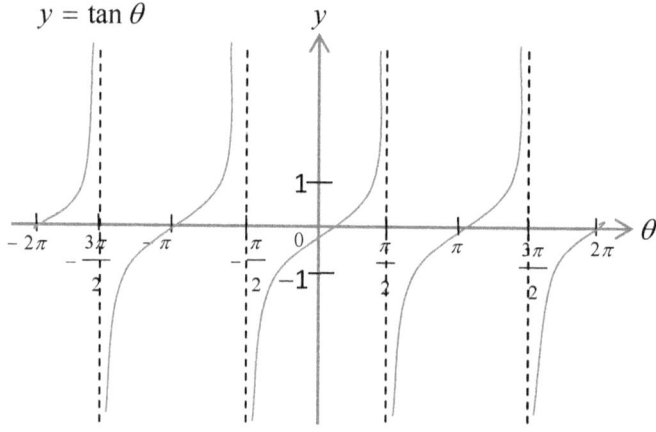

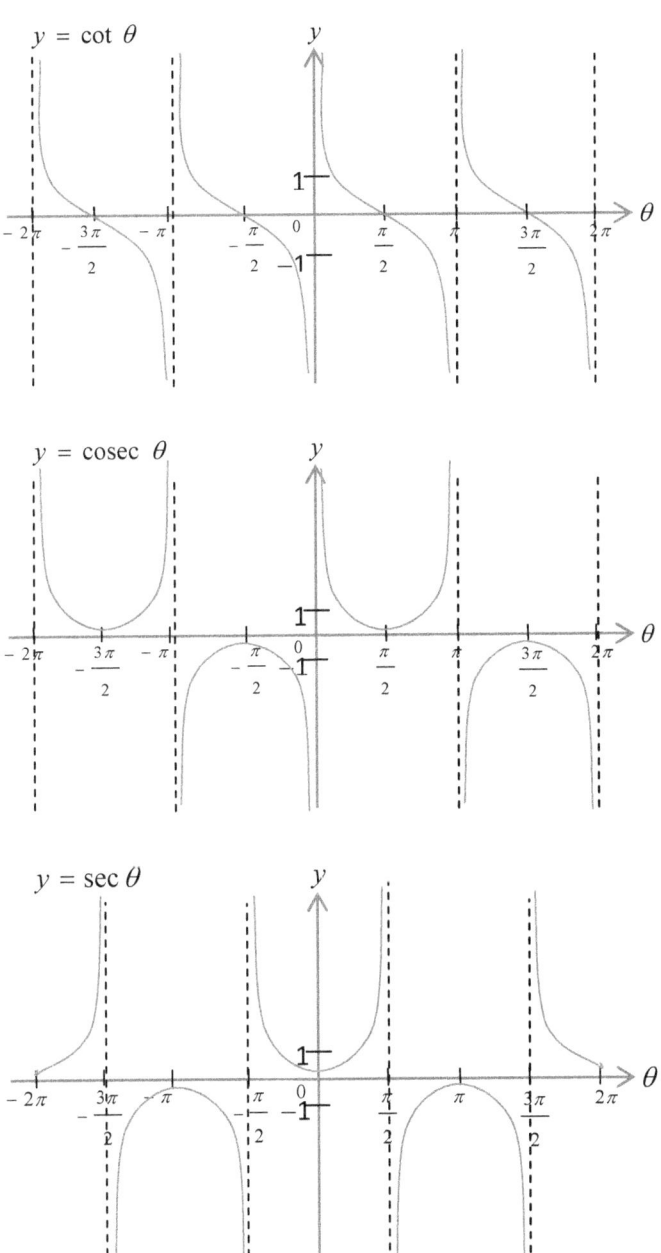

Simple Transformation of Trigonometric Curves

The following shows the graph of $f(x) = \sin x$ and four of its transformations. Each curve is labeled. Study them carefully with reference to the section, Transformations of Graphs of Functions (Topic1).

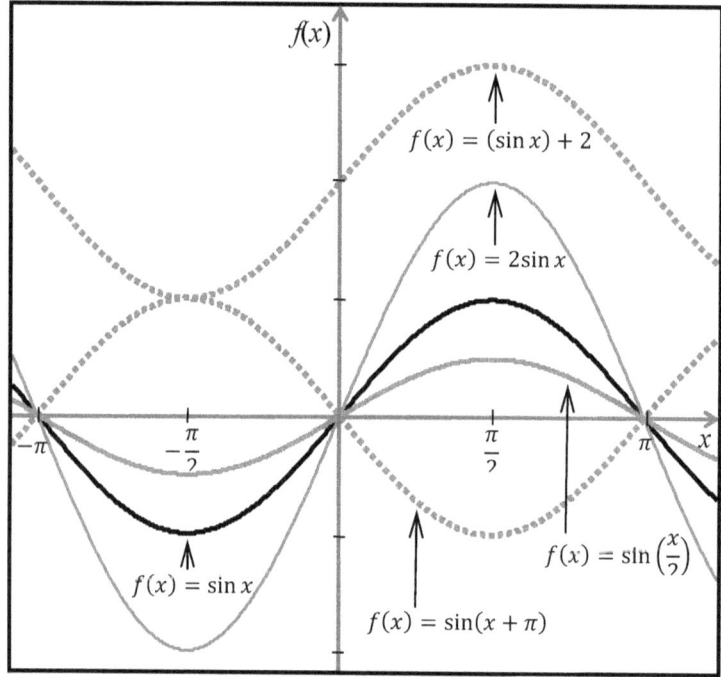

Now:

(a) Draw the graph of $f(x) = \cos x$ and as many of its transformations as possible.

(b) Draw the graph of $f(x) = \tan x$ and as many of its transformations as possible.

The Sine and Cosine Formulae

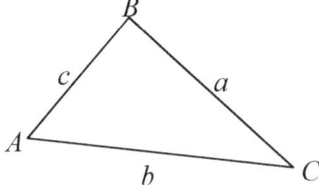

Sine Formulae

$$\frac{\sin A}{a} = \frac{\sin B}{b} = \frac{\sin C}{c}$$

$$\frac{a}{\sin A} = \frac{b}{\sin B} = \frac{c}{\sin C}$$

Cosine Formulae

$$a^2 = b^2 + c^2 - 2bc \cos A$$

$$b^2 = a^2 + c^2 - 2ac \cos B$$

$$c^2 = a^2 + b^2 - 2ab \cos C$$

The sine formulae or sine rule is used when
(a) 2 angles and 1 side of a triangle are given.
(b) 2 sides and a non-included angle of a triangle are given.

The cosine formulae or cosine rule is used when
(a) 2 sides and an included angle of a triangle are given.
(b) 3 sides of a triangle are given.

Area of $\triangle ABC$

Area of $\triangle ABC$, $S = \dfrac{1}{2}ab\sin C = \dfrac{1}{2}ac\sin B = \dfrac{1}{2}bc\sin A$

Examples

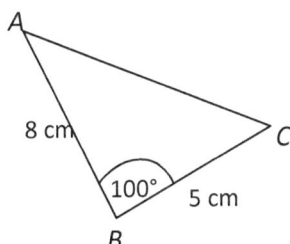

In the figure on the left, calculate
(i) AC
(ii) Angle BAC
(iii) Angle BAC
(iv) The area of the triangle ABC.

Solution

(i) By the cosine rule, $b^2 = a^2 + b^2 - 2ac \cos B$

$\Rightarrow b^2 = 5^2 + 8^2 - 2(5)(8) \cos 100°$

$\Rightarrow b = \sqrt{102.89} = 10.14$ cm

(ii) By the sine rule, $\dfrac{\sin A}{5} = \dfrac{\sin 100°}{10.14}$

$\sin A = \dfrac{5\sin 100°}{10.14} = 0.4856 \Rightarrow$ angle $BAC = 29.1°$

(iii) $\angle ACB = 180 - (100 + 29.1°) = 50.9°$

(iv) The area of the triangle $ABC = \dfrac{1}{2}ac \sin B$

$= \dfrac{1}{2}(5)(8)\sin 100° = 19.7$ cm^2

Inverse Trigonometric Functions

$\sin^{-1} x = \theta \Leftrightarrow -90° \le \theta \le 90°$

$\cos^{-1} x = \theta \Leftrightarrow -0° \le \theta \le 180°$

$\tan^{-1} x = \theta \Leftrightarrow -90° \le \theta \le 90°$

Small Angles

As $\theta \to 0$, $\sin \theta = \theta \Leftrightarrow \dfrac{\sin \theta}{\theta} = 1$ or $\dfrac{\theta}{\sin \theta} = 1$

$\cos \theta \approx 1 - \dfrac{1}{2}\theta^2$

$\tan \theta = \theta \Rightarrow \dfrac{\tan \theta}{\theta} = 1$ or $\dfrac{\theta}{\tan \theta} = 1$

TOPIC 7
LIMITS AND DIFFERENTIATION

Limits

The limit of $f(x)$ as x tends to a is written as $\lim\limits_{x \to a} f(x) = L$.

Laws of Limits

1. Limits of $\frac{1}{x}$

$$\lim\limits_{x \to \infty}\left(\frac{1}{x}\right) = 0$$

2. Limits of a Constant

$$\lim\limits_{x \to a}(k) = k$$

3. Limits of a Linear Expression

$$\lim\limits_{x \to a}(mx + b) = ma + b$$

4. Limits of a Sum

$$\lim\limits_{x \to a}\left(f(x) + g(x)\right) = \lim\limits_{x \to a} f(x) + \lim\limits_{x \to a} g(x)$$

5. Limits of a Product

$$\lim\limits_{x \to a}\left(f(x) \cdot g(x)\right) = \lim\limits_{x \to a} f(x) \cdot \lim\limits_{x \to a} g(x)$$

6. Limits of a Quotient

$$\lim\limits_{x \to a}\left(\frac{f(x)}{g(x)}\right) = \frac{\lim\limits_{x \to a} f(x)}{\lim\limits_{x \to a} g(x)}, \text{ where } \lim\limits_{x \to a} g(x) \neq 0$$

7. Limits of a Root

$$\lim\limits_{x \to a} \sqrt[n]{f(x)} = \sqrt[n]{\lim\limits_{x \to a} f(x)}$$

Gradient of a Straight Line $m = \dfrac{y_2 - y_1}{x_2 - x_1}$

Gradient of a Curve at Any Point

The gradient of a curve at a point is defined as

$$f'(x) = \frac{dy}{dx} = \lim_{x \to h} \frac{f(x+h) - f(x)}{h}$$

This formula is used to differentiate functions from first principle.

Function of a Function and Implicit Functions

The chain rule is used to differentiate a function of a function and implicit functions.
The chain rule states that if y is a function of a function i.e.

$y = g(v)$ and $v = f(x)$ then $\dfrac{dy}{dx} = \dfrac{dy}{dv} \cdot \dfrac{dv}{dx}$.

Parametric Functions

$y = g(t)$ and $x = f(t) \Rightarrow \dfrac{dy}{dx} = \dfrac{g'(t)}{f'(t)}$

The Second Derivative

The differential of $\dfrac{dy}{dx}$ is called the second derivative and is

denoted by $f''(x)$ or $\dfrac{d^2 y}{dx^2}$.

Tangents and Normals

The gradient $\dfrac{dy}{dx}$ of a curve at a point is equal to the gradient m of the tangent at that point. The equation of a tangent to a curve at a point (x_1, y_1) is $y = mx + (y_1 - mx_1)$, the gradient of the normal to the curve at the point is $-\dfrac{1}{m}$ and its equation is given by $\dfrac{y-y_1}{x-x_1} = -\dfrac{1}{m}$ which we can rearrange to have

$$y = -\dfrac{1}{m}x + \left(y_1 + \dfrac{1}{m}x_1\right).$$

At the point (x_1, y_1), the equation of the tangent to the second degree curve $f(x, y) = ax^2 + by^2 + cxy + dx + ey + k$ is

$$g(x, y) = axx_1 + byy_1 + \dfrac{c}{2}(xy_1 + x_1 y) + \dfrac{d}{2}(x + x_1) + \dfrac{e}{2}(y + y_1) + k$$

Stationary Points (Maximum and Minimum Turning Points and Points of Inflexion)

A turning point is a point at which a curve changes direction. There are three types of turning points distinguished by their values of $\dfrac{d^2 y}{dx^2}$ as shown in the figure below.

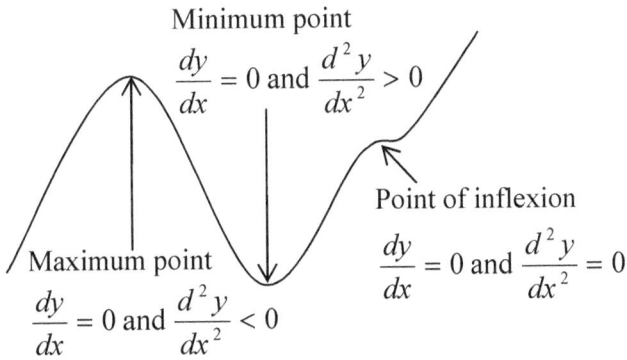

Minimum point
$\dfrac{dy}{dx} = 0$ and $\dfrac{d^2 y}{dx^2} > 0$

Point of inflexion
$\dfrac{dy}{dx} = 0$ and $\dfrac{d^2 y}{dx^2} = 0$

Maximum point
$\dfrac{dy}{dx} = 0$ and $\dfrac{d^2 y}{dx^2} < 0$

Standard Differentials

Type	$f(x)$	$f'(x)$ or $\dfrac{dy}{x}$
Constant	c	0
Monomials	x^n	nx^{n-1}
	ax^n	anx^{n-1}
Exponential	e^x	e^x
	e^{ax}	ae^{ax}
	e^{ax+b}	ae^{ax+b}
	a^x	$a^x \ln a$
	x^x	$x^x(1+\ln x)$
Logarithmic	$\ln x$	$\dfrac{1}{x}$
	$\ln f(x)$	$\dfrac{1}{f(x)}\dfrac{dy}{dx}$
	$\ln ax$	$\dfrac{1}{x}\quad \left[\text{not } \dfrac{a}{x}\right]$
	$\ln(ax+b)$	$\dfrac{a}{ax+b}$
Trigonometric	$\sin x$	$\cos x$
	$\cos x$	$-\sin x$
	$\tan x$	$\sec^2 x$
	$\cot x$	$-\cosec^2 x$
	$\sec x$	$\sec x \tan x$
	$\cosec x$	$-\cosec x \cot x$
	$\sin ax$	$a\cos x$
	$\cos ax$	$-a\sin ax$
	$\tan ax$	$a\sec^2 ax$
	$\cot ax$	$a\cosec^2 ax$
	$\sec ax$	$a\sec ax \tan ax$
	$\cosec ax$	$-a\cosec ax \cot ax$
	$\sin(ax+b)$	$a\cos(ax+b)$
	$\cos(ax+b)$	$-a\sin(ax+b)$

Type	$f(x)$	$f'(x)$ or $\dfrac{dy}{x}$
	$\tan(ax+b)$	$a\sec^2(ax+b)$
	$\cot(ax+b)$	$a\operatorname{cosec}^2(ax+b)$
	$\sec(ax+b)$	$a\sec(ax+b)\tan(ax+b)$
	$\operatorname{cosec}(ax+b)$	$-a\operatorname{cosec}(ax+b)\cot(ax+b)$
Inverse Trigonometric (*To differentiate or integrate trigonometric functions, the angles must be in radians.*)	$\sin^{-1}x$	$\dfrac{1}{\sqrt{1-x^2}}$
	$\cos^{-1}x$	$-\dfrac{1}{\sqrt{1-x^2}}$
	$\tan^{-1}x$	$\dfrac{1}{x^2+1}$
	$\cot^{-1}x$	$-\dfrac{1}{x^2+1}$
	$\sec^{-1}x$	$\dfrac{1}{x\sqrt{x^2-1}}$
	$\operatorname{cosec}^{-1}x$	$-\dfrac{1}{x\sqrt{x^2-1}}$
	$\sin^{-1}ax$	$\dfrac{a}{\sqrt{1-a^2x^2}}$
	$\cos^{-1}ax$	$-\dfrac{1}{\sqrt{1-a^2x^2}}$
	$\tan^{-1}ax$	$\dfrac{a}{a^2x^2+1}$
	$\cot^{-1}ax$	$-\dfrac{a}{a^2x^2+1}$
	$\sec^{-1}ax$	$\dfrac{1}{x\sqrt{a^2x^2-1}}$
	$\operatorname{cosec}^{-1}ax$	$-\dfrac{1}{x\sqrt{a^2x^2-1}}$

GENERAL CURVE SKETCHING

To sketch a curve $y = f(x)$, we may use some or all of the following to analyze the curve.

1. **Intercepts**

   To find the intercepts with the x-axis, substitute $y = 0$ in the equation of the curve and solve for x. To find the intercepts with the y-axis, substitute $x = 0$ in the equation of the curve and solve for y.

2. **Stationary Points**

   To find the stationary point(s), find the value of x for which $\frac{dy}{dx} = 0$. Determine the value of $\frac{d^2y}{dx^2}$ for this value of x. The following table summarizes the values of $\frac{dy}{dx}$ and $\frac{d^2y}{dx^2}$ and the nature of the curve at the different stationary (or turning) points and closed to the stationary (or turning) points.

maximum	minimum	Point of inflexion
$\frac{dy}{dx} = 0$	$\frac{dy}{dx} = 0$	$\frac{dy}{dx} = 0$
$\frac{d^2y}{dx^2} \leq 0$	$\frac{d^2y}{dx^2} \geq 0$	$\frac{d^2y}{dx^2} = 0$

3. **Asymptotes**

   A vertical asymptote is the straight line $y = a$, at which as $x \to \pm\infty$, the curve continues to approach but never touching the line $y = a$.

   A horizontal asymptote is the straight line $x = a$, at which as $y \to \pm\infty$, the curve continues to approach but never touching the line $x = a$.

   A slant asymptote is the straight line $y = mx + c$, at which the curve continues to approach but never touching the line $y = mx + c$.

4. **Empty regions**

   Empty regions are the regions where no part of the graph lies. To find empty regions we determine the sign of the function $y = f(x)$ in the range determined by the critical values of x. When the function is a proper rational function with a quadratic denominator, we determine the set of values of y for which x is real.

5. **Graphs of Parametric functions**

   To sketch the graph of parametric functions $x = f(t)$ and $y = g(t)$, we may eliminate t from the parametric equations to have the equation in the form $y = h(x)$. However, in most cases it is not always easy to eliminate t. In such cases, we make a table of values of x and y for different values of t. Then use this table of values to draw the graph of y against x.

Examples

Sketch the curves

(a) $2y = \dfrac{3x-4}{x(x-1)}$

(b) $x = t - \sin t, y = 1 - \cos t$ for $-2\pi \le t \le \pi$.

Solution

(i)   **Intercepts**

When $y = 0, x = \dfrac{4}{3}$ and when $x = 0$, there is no definite value of y. Therefore, the point $\left(\dfrac{4}{3}, 0\right)$ is the intercept with the x-axis.

(ii)   Asymptotes

As $x \to \pm\infty$, $y \to 0$. Therefore $y = 0$ is a horizontal asymptote. As $y \to \pm\infty$, $x \to 0$ or $x \to 1$. Therefore, $x = 0$ and $x = 1$ are vertical asymptotes.

(iii)   Stationary Points

Rearranging the equation of the curve,

$2yx^2 - (2y + 3)x + 4 = 0$.

For real values of x, $b^2 - 4ac \geq 0$.

$\Rightarrow (2y + 3)^2 - 4(2y)(4) \geq 0$

$$4y^2 - 20y + 9 \geq 0$$

$$(2y - 9)(2y - 1) \geq 0$$

$$\Rightarrow y \leq \frac{1}{2} \text{ or } y \geq \frac{9}{2}$$

Therefore, y does not exist in the range $\frac{1}{2} < y < \frac{9}{2}$.

The point where the curve touches the boundary lines $y = \frac{1}{2}$ and $y = \frac{9}{2}$ are the values of x for which $2yx^2 - (2y + 3)x + 4 = 0$ has equal roots. Recall from topic 1 that at this point $x = -\frac{b}{2a}$. i.e. $x = \frac{2y+3}{4y}$.

Substituting $y = \frac{1}{2}$ and $y = \frac{9}{2}$ in $x = \frac{2y+3}{4y}$, we see that $\left(2, \frac{1}{2}\right)$ is the maximum point and $\left(\frac{2}{3}, \frac{9}{2}\right)$ is the minimum point.

(iv)   Empty regions

The critical values are $x = -3$, $x = -1$ and $x = 2$.

	$x < 0$	$0 < x < 1$	$1 < x < \dfrac{4}{3}$
x	$-$	$+$	$+$
$x - 1$	$-$	$-$	$+$
$3x - 4$	$-$	$-$	$-$
y	$-$	$+$	$-$

We can now sketch the curve.

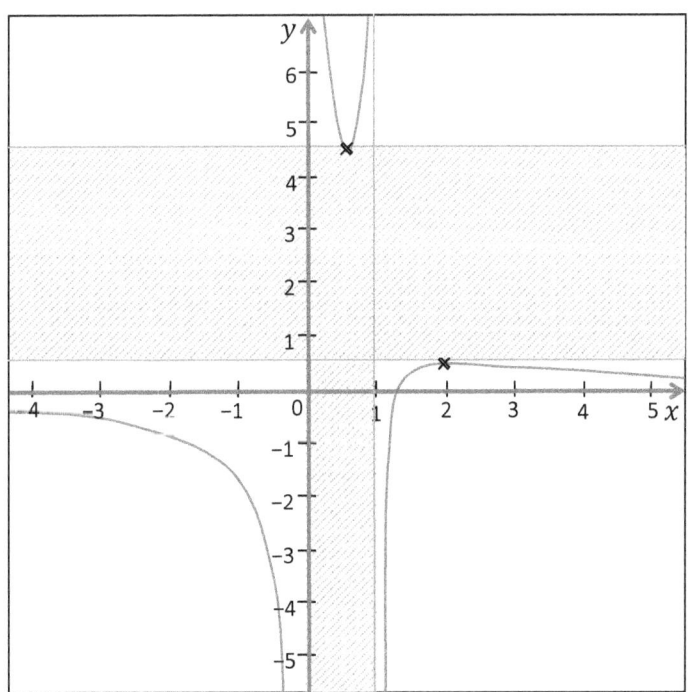

(b) We make a table of values.

t	-2π	$-\dfrac{5\pi}{3}$	$-\dfrac{3\pi}{2}$	$-\dfrac{4\pi}{3}$	$-\pi$	$-\dfrac{2\pi}{3}$	$-\dfrac{\pi}{2}$	$-\dfrac{\pi}{3}$
x	-6.3	-6.1	-5.7	-5.1	-3.1	-1.2	-0.7	-0.2
y	0	0.5	1	1.5	2	1.5	1	0.5

51

0	$\dfrac{\pi}{3}$	$\dfrac{\pi}{2}$	$\dfrac{2\pi}{3}$	π	$\dfrac{4\pi}{3}$	$\dfrac{3\pi}{2}$	$\dfrac{5\pi}{3}$	2π
0	0.2	0.6	1.2	3.1	5.1	5.7	6.1	6.3
0	0.5	1	1.5	2	1.5	1	0.5	0

We now sketch the curve.

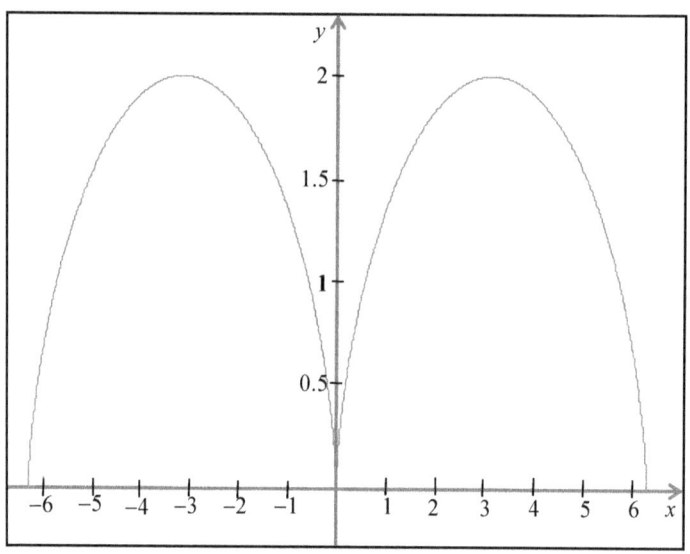

TOPIC 8
INTEGRATION

Integration by Recognition

Much of integration can be done by recognizing a function $f(x)$ as the derivative $f'(x)$ of the function $f(x)$. Therefore, we can simply reverse the table of standard differentials in topic 7 to obtain a table of standard integrals as follows.

Standard Integrals

Let a, b and c be constants and $n \in \mathbb{R}$. The constant of integration has been voluntarily omitted and the reader should remember to insert it in all indefinite integration.

Type	$f(x)$	$f'(x)$ or $\dfrac{dy}{dx}$
Constant	c	cx
Polynomials	$ax^n,\ n \neq -1$	$\dfrac{ax^{(n+1)}}{(n+1)}$
	$(ax+b)^n, n \neq -1$	$\dfrac{(ax+b)^{n+1}}{a(n+1)}$
Exponential	e^x	e^x
	e^{ax+b}	$\dfrac{1}{a}e^{ax+b}$
	$f'(x)e^{f(x)}$	$e^{f(x)}$
	a^x	$\dfrac{1}{\ln a}a^x$

Type	$f(x)$	$f'(x)$ or $\dfrac{dy}{dx}$
	a^{bx+c}	$\dfrac{1}{b\ln a}a^{bx+c}$
	$\sin x$	$-\cos x$
	$\cos x$	$\sin x$
	$\tan x$	$-\ln|\cos x|$
Trigonometric	$\cot x$	$\ln|\sin x|$
(*To differentiate*	$\sin(ax+b)$	$-\dfrac{1}{a}\cos(ax+b)$
or integrate		
trigonometric	$\cos(ax+b)$	$-\dfrac{1}{a}\sin(ax+b)$
functions, the		
angles must be	$\tan(ax+b)$	$-\dfrac{1}{a}\ln\left|\cos(ax+b)\right|$
in radians.)	$\cot(ax+b)$	$\dfrac{1}{a}\ln|\sin(ax)|$
	$\cos x\sin^n x$	$\dfrac{1}{n+1}\sin^{(n+1)}x$
	$\sin x\cos^n x$	$-\dfrac{1}{n+1}\cos^{(n+1)}x$
	$\sec^2 x\tan^n x$	$\dfrac{1}{n+1}\tan^{(n+1)}x$
Logarithmic	$\ln x$	$x(\ln x-1)$
	$\ln ax$	$x\ln ax-x$

Type	$f(x)$	$f'(x)$ or $\dfrac{dy}{dx}$
	$\ln(ax+b)$	$\dfrac{(ax+b)}{a}\big(\ln(ax+b)-1\big)$
	$\dfrac{1}{x}$	$\ln|x|$
	$\dfrac{1}{ax+b}$	$\dfrac{1}{a}\ln|ax+b|$
	$\dfrac{f'(x)}{f(x)}$	$\ln|f(x)|$
	$\dfrac{1}{\sqrt{1-x^2}}$	$\sin^{-1}x$
Rational	$\dfrac{1}{\sqrt{1-a^2x^2}}$	$\dfrac{1}{a}\sin^{-1}ax$
	$\dfrac{1}{x^2+1}$	$\tan^{-1}x$
	$\dfrac{1}{a^2x^2+1}$	$\dfrac{1}{a}\tan^{-1}ax$
	$\dfrac{1}{x\sqrt{x^2-1}}$	$\sec^{-1}x$
	$\dfrac{1}{x\sqrt{a^2x^2-1}}$	$\sec^{-1}ax$

Integration by Substitution

Example
Evaluate $\int (2x+4)^5\,dx$

Let $u = 2x+4 \implies \dfrac{du}{dx} = 2$ and $dx = \dfrac{du}{2}$

$$\therefore \int (2x+4)^5 dx = \int u^5 \frac{du}{2} = \left(\frac{1}{2}\right)\left(\frac{u^6}{6}\right) = \frac{(2x+4)^6}{12} + k.$$

The following substitutions are worth noting.

Integral	Substitution
$\int (ax+b)^n \, dx$	
$\int \dfrac{1}{(ax+b)^n} dx$	$u = ax+b$
$\int \text{Trig}\,(ax+b)\,dx$	
$\int \dfrac{x^n}{a+x^{n+1}} dx$	$u = a + x^{n+1}$
$\int \left(x\sqrt{ax+b}\right) dx$	$u = \sqrt{ax+b}$
$\int \dfrac{x^n}{\sqrt{a+x^{n+1}}} dx$	$u = \sqrt{a+x^{n+1}}$
$\int \dfrac{1}{\sin x \cos x} dx$	$u = \tan x$
$\int \dfrac{1}{\sqrt{1-x^2}} dx$	$x = \sin\theta$
$\int \dfrac{1}{x^2+1} dx$	$x = \tan\theta$
$\int \dfrac{1}{1+\sin x} dx$	$t = \tan \dfrac{\theta}{2}$
$\int \dfrac{1}{b+a\cos 2x} dx$	$t = \tan\theta$
$\int \dfrac{1}{x^2+a^2} dx$	$x = au$

Integration of a^x w.r.t. x

Let $y = a^x \Rightarrow \ln y = x \ln a$ and $y = e^{x \ln a}$ or $a^x = e^{x \ln a}$.

$$\therefore \int a^x dx = \int e^{x \ln a} dx = \frac{e^{x \ln a}}{\ln a} + k.$$

Integration of Rational Expressions (Fractions)

(a) In $\int \frac{f'(x)}{f(x)} dx$ the numerator is basically the derivative of the denominator.

Let $y = f(x) \Rightarrow \frac{dy}{dx} = f'(x)$ and $dx = \frac{dy}{f'(x)}$.

$$\Rightarrow \int \frac{f'(x)}{f(x)} dx = \int \frac{f'(x)}{y} \cdot \frac{dy}{f'(x)} = \int \frac{1}{y} dy = \ln|y| + k.$$

$$\therefore \int \frac{f'(x)}{f(x)} dx = \ln|f(x)| + k.$$

Example

Evaluate $\int \frac{\cos x}{1+\sin x} dx$.

$$\frac{d(1 + \sin x)}{dx} = \cos x \Rightarrow \int \frac{\cos x}{1 + \sin x} dx = \ln|1 + \sin x| + k$$

(b) If the numerator is the derivative of some function of the denominator change the variable.

Example

Integrate $\frac{2x}{\sqrt{x^2+1}}$ w.r.t. x.

Let $y = x^2 + 1 \Rightarrow \frac{dy}{dx} = 2x \Rightarrow dx = \frac{dy}{2x}$.

$$\int \frac{2x}{\sqrt{x^2 + 1}} dx = \int \frac{2x}{\sqrt{y}} \cdot \frac{dy}{2x} = \int \frac{1}{\sqrt{y}} dy = \int y^{-\frac{1}{2}} dy$$

$$\int \frac{2x}{\sqrt{x^2 + 1}} dx = 2y^{\frac{1}{2}} + k = 2\sqrt{x^2 + 1} + k$$

(c) **Integration Using Partial Fractions**

Go back and revise topic 5.

57

To integrate rational functions by using partial fractions, first decompose the function into partial fractions and use any of the following rules.

(i)   For a term with linear denominator,

$$\int \frac{1}{ax+b}\, dx = \frac{1}{a}\ln|ax+b| + k$$

(ii)   For a term with a repeated factor denominator,

$$\int \frac{1}{(ax+b)^2}\, dx = -\frac{1}{a(ax+b)} + k$$

(iii)   For a term with non-factorable quadratic denominator, complete the square of the quadratic to have it in the form $(x+a)^2 + b^2$ and then use the substitution $u = x + a$ to integrate the function.

Example

Evaluate $\int \frac{3x}{(x-1)(x-2)}\, dx$.

$$\int \frac{3x}{(x-1)(x-2)}\, dx = \int \frac{6}{(x-2)}\, dx - \int \frac{3}{(x-1)}\, dx$$

$$= 6\ln|x-2| - 3\ln|x-1| + k$$

Integration of Products

(a) Product of the form $f'(x).e^{f(x)}$

$$\frac{d\left(e^{f(x)}\right)}{dx} = f'(x).e^{f(x)} \implies \int f'(x).e^{f(x)}\, dx = e^{f(x)} + k$$

Example

Integrate $x^2 e^{x^3}$ w.r.t. x.

$$\int x^2 e^{x^3}\, dx = \frac{1}{3}e^{x^3} + k$$

(b) One Factor is a Function of the derivative of the other

In this case use a substitution to change the variable.

Example

Evaluate (i) $\int x^2\sqrt{x^3+5}\,dx$ (ii) $\int \cos x \sin^3 x\,dx$

(i) Let $u = x^3 + 5 \Longrightarrow \frac{du}{dx} = 3x^2 \Longrightarrow dx = \frac{du}{3x^2}$

$\therefore \int x^2\sqrt{x^3+5}\,dx = \int x^2\sqrt{u}\,\frac{du}{3x^2} = \frac{1}{3}\int u^{\frac{1}{2}}du$

$$\int x^2\sqrt{x^3+5}\,dx = \frac{1}{3}\left(\frac{u^{\frac{3}{2}}}{\frac{3}{2}}\right) + k = \frac{2}{9}(x^3+5)^{\frac{3}{2}} + k$$

(ii) Let $u = \sin x \Longrightarrow dx = \frac{du}{\cos x}$

$\Longrightarrow \int \cos x \sin^3 x\,dx = \int \cos x\, u^3\,\frac{du}{\cos x} = \int u^3\,du = \frac{u^4}{4} + k$

$\therefore \int \cos x \sin^3 x\,dx = \frac{\sin^4 x}{4} + k.$

Generally

$$\int \cos x \sin^n x\,dx = \frac{\sin^{(n+1)}x}{(n+1)} + k \text{ and }$$

$$\int \sin x \cos^n x\,dx = -\frac{\cos^{(n+1)}x}{(n+1)} + k$$

(c) Integration by Parts

We can use the chain rule show that $\int v\,du = uv - \int u\,dv$.

We call the process of using this formula to integrate integration by parts and we must choose v and $\frac{du}{dx}$ in such a way that it is easier to integrate $\int u\,dv$ than $\int v\,du$.

Example

Evaluate (i) $\int xe^x dx$. (ii) $\int \ln x\,dx$

$$\int v\,du = uv - \int u\,dv$$

(i) Let $v = x \Longrightarrow dv = dx$ and $du = e^x dx \Longrightarrow u = e^x$

$$\Longrightarrow \int xe^x dx = xe^x - \int e^x\,dx$$

$$\Rightarrow \int xe^x dx = xe^x - e^x + k = (x-1)e^x + k$$

(ii) $\int \ln x \, dx = \int (1) \ln x \, dx$

Let $v = \ln x \Rightarrow dv = \left(\frac{1}{x}\right) dx$ and $du = (1)dx \Rightarrow u = x$

$$\Rightarrow \int \ln x \, dx = x \ln x - \int (1) dx$$

$$= x \ln x - x + k = (x-1) \ln x$$

Sometimes we have to apply the formula more than once before obtaining the integral like in $\int e^x \cos x dx$. Try this yourself. The answer is

$$\int e^x \cos x dx = \frac{1}{2} e^x (\sin x + \cos x) + k$$

Integration of a Product of Trigonometric Ratios

Use the factor formulae to convert the product of multiple angle trigonometric ratios to a sum of trigonometric ratios.

Example

Evaluate $\int \sin 5\theta \cos 3\theta d\theta$.

$$\sin A \cos B = \frac{1}{2}\sin(A+B) + \frac{1}{2}\sin(A-B).$$

$$\Rightarrow \int \sin 5\theta \cos 3\theta d\theta = \frac{1}{2}\int (\sin 8\theta + \sin 2\theta) d\theta.$$

$$= \frac{1}{16}\cos 8\theta - \frac{1}{4}\cos 2\theta + k$$

Integration of a Powers of Trigonometric Ratios

For the functions $y = \sin^n x$ and $y = \cos^n x$,

(1) If n is odd, use the Pythagorean identity
$\sin^2 x + \cos^2 x = 1$ to convert the function into a form that can be integrated.

(2) If n is even, write the function in the forms

$$y = \left(\sin^2 x\right)^n \text{ or } y = \left(\cos^2 x\right)^n \text{ and use the double}$$

angle identities as many times as possible to convert the function into a form that can be integrated.

Example

Evaluate $\int \sin^5 \theta \, d\theta$

$$\sin^5 \theta = \sin \theta \, (\sin^2 \theta)^2 = \sin \theta \, (1 - \cos^2 \theta)^2$$
$$= \sin \theta \, (1 - 2\cos^2 \theta + \cos^4 \theta)$$
$$\therefore \int \sin^5 \theta \, d\theta = \int (\sin \theta - 2 \sin \theta \cos^2 \theta + \sin \theta \cos^4 \theta) d\theta$$

$$= -\cos \theta - \frac{2}{3}\cos^2\theta - \frac{1}{5}\cos^5\theta + k$$

Definite Integral

Example

Evaluate (i) $\int_0^{\frac{\pi}{3}} \cos \theta \, d\theta$ (ii) $\int_1^4 \frac{1}{(x+3)^2} \, dx$

(i) $\int_0^{\frac{\pi}{2}} \cos \theta \, d\theta = \sin \theta]_0^{\frac{\pi}{2}} = \sin\frac{\pi}{2} - \sin 0 = 1$

(ii) $\int_1^4 \frac{1}{(x+3)^2} \, dx = \int_1^4 (x + 3)^{-2} dx$

$$= -\frac{1}{(x+3)}\Big]_1^4 = -\frac{1}{7} + \frac{1}{2} = \frac{5}{14}$$

Example

Find the area bounded by, the $x -$ axis, the lines $x = a$ and $x = b$ and the curve $y = 3x^2$.

$$A = \int_a^b 3x^2 dx = x^3]_a^b = b^3 - a^3$$

TOPIC 9
FIRST ORDER
DIFFERENTIAL EQUATIONS

Solving First Order Differential Equations

To solve a first order differential equation, collect the terms in x on the same side with dx and the terms in y on the same side as dy and then integrate both sides. The following shows how the various forms should be treated.

1. $\dfrac{dy}{dx} = f(x) \Rightarrow \int dy = \int f(x)dx + k$

2. $\dfrac{dy}{dx} = f(y) \Rightarrow \int \dfrac{1}{f(y)} dy = \int dx + k$

3. $\dfrac{dy}{dx} = f(y)g(x) \Rightarrow \int \dfrac{1}{f(y)} dy = \int g(x)dx + k$

4. $\dfrac{dy}{dx} = \dfrac{f(y)}{g(x)} \Rightarrow \int \dfrac{1}{f(y)} dy = \int \dfrac{1}{f(x)} dx + k$

5. $f(y)\dfrac{dy}{dx} = g(x) \Rightarrow \int f(y)dy = \int g(x)dx + k$

Example

Find y in terms of x given that $(1 + x)\dfrac{dy}{dx} = (1 - x)y$.

Solution

$$\int \dfrac{1}{y} dy = \int \left(\dfrac{1 - x}{1 + x}\right) dx \Rightarrow \ln y = \int \left(\dfrac{2}{1 + x} - 1\right) dx$$

$$\ln y = 2\ln(1 + x) + k \Rightarrow y = e^k (1 + x)^2$$

We call this the general solution of a differential equation. The general solution constitutes a family of curves. If some parameters are given the integration constant can be found and substituted to obtain the particular solution.

Example

Solve in the form $y = f(x)$, the differential equation

$(x + 1)\frac{dy}{dx} = 1 - y$, given that $y = 3$ when $x = 0$.

Solution

$$(x + 1)\frac{dy}{dx} = 1 - y \Rightarrow -\int \frac{1}{y - 1}\,dy = \int \frac{1}{x + 1}\,dx$$

$$\Rightarrow -\ln(y - 1) = \ln(x + 1) + k$$

Substituting $y = 3$ and $x = 0$,

$$\Rightarrow -\ln(2) = \ln(1) + k \Rightarrow k = -\ln(2)$$

$$\Rightarrow -\ln(y - 1) = \ln(x + 1) + -\ln(2) \Rightarrow \frac{1}{y - 1} = \frac{x + 1}{2}.$$

Rearranging, we have $y = \frac{x+3}{x+1}$.

Formulation of First Order Differential Equations

If the rate of increase of a quantity Q with respect to another quantity θ is proportional to the value of Q, then $\dfrac{dQ}{d\theta} = kQ$,

where k is a constant.

$$\therefore \int \frac{1}{Q}\,dQ = k\int d\theta \Rightarrow \ln AQ = k\theta \Rightarrow Q = Be^{k\theta}$$

If the rate of decrease of a quantity Q with respect to another quantity θ is proportional to the value of Q, then

$$-\frac{dQ}{d\theta} = kQ, \text{ where } k \text{ is a constant.}$$

$$\therefore \int \frac{1}{Q}\,dQ = -k\int d\theta \Rightarrow \ln AQ = -k\theta \Rightarrow Q = Be^{-k\theta}$$

TOPIC 10
SEQUENCE AND SERIES

Definition of a Number Sequence

A number sequence is an ordered arrangement of numbers called terms using a specific rule. The terms in a sequence are usually denoted by $u_1, u_2, u_3, \cdots u_r, \cdots$ where $r \in \mathbb{Z}^+$ and u_r is the rth term of the sequence. Sometimes, the first term is denoted by a.

Definition of a Number Series

A series is generated by adding the terms of a sequence. Thus, if $u_1, u_2, \cdots u_r, \cdots$ is a sequence, then $u_1 + u_2 + \cdots + u_r + \cdots$ is the corresponding series.

The sum of first n terms of a series is denoted by $S_n = \sum\limits_{r=1}^{n} u_r$.

Finite and Infinite Sequences and Series

A finite sequence or series is one that contains a definite number of terms.
An infinite sequence or series is one that continues indefinitely.

Arithmetic Progression (A.P.)

For a series to be an A.P., $u_{n+1} - u_n = u_n - u_{n-1} = d \quad \forall n \in \mathbb{Z}^+$,

where d is a constant called the common difference of the A.P.

The nth term of an A.P. is given by
$$U_n = a + (n - 1)d.$$

The sum of n terms of an A.P. is given by
$$S_n = \frac{n}{2}\{2a + (n - 1)d\}$$

Arithmetic Mean of a set of n numbers is given by $\bar{x} = \dfrac{\sum\limits_{r=1}^{n} u_r}{n}$.

64

If $u_{n-1}, u_n,$ and u_{n+1} are consecutive terms of an A.P., then

$$u_n = \frac{u_{n-1} + u_{n+1}}{2}.$$

Geometric Progression (G.P.)

For a series to be a G.P., $\dfrac{u_{n+1}}{u_n} = \dfrac{u_n}{u_{n-1}} = r$, $\forall n \in \mathbb{Z}^+$, where r is a constant called the common ratio of the G.P. The nth term of a G.P. is given by $u_n = ar^{n-1}$

The sum of n terms of a G.P. is given by

$$S_n = \frac{a(r^n - 1)}{r - 1} \text{ for } |r| > 1 \text{ or } S_n = \frac{a(1 - r^n)}{1 - r} \text{ for } |r| < 1$$

The geometric mean of a set of n numbers $u_1, u_2, u_3, \cdots u_n$ is given by $G.M. = \sqrt[n]{(u_1)(u_2)(u_3) \cdots (u_n)}$.

If $u_{n-1}, u_n,$ and u_{n+1} are consecutive terms of a G.P., then

$$\frac{u_n}{u_{n-1}} = \frac{u_{n+1}}{u_n} \Rightarrow u_n = \sqrt{(u_{n+1})(u_{n-1})}.$$

Convergent Geometric Series

The condition for a G.P. $u_1, u_2, u_3 \ldots, u_n$ to be convergent is that $|r| < 1$. If an infinite geometric series is convergent then the sum of its terms called the sum to infinity is given by $S_\infty = \dfrac{a}{1 - r}$

Summation of Series

Given that a is a constant then,

1. $\displaystyle\sum_{r=1}^{n} a = an$

2. $\displaystyle\sum_{r=1}^{n} af(r) = a\sum_{r=1}^{n} f(r)$

3. $\displaystyle\sum_{r=1}^{n} [f(r) + g(r)] = \sum_{r=1}^{n} f(r) + \sum_{r=1}^{n} g(r)$

Summation of Some Standard Series

1. $\displaystyle\sum_{r=1}^{n} 1 = n$

2. $\displaystyle\sum_{r=1}^{n} r = \frac{n}{2}(n+1)$

3. $\displaystyle\sum_{r=1}^{n} r(r+1) = \frac{n}{3}(n+1)(n+2)$

4. $\displaystyle\sum_{r=1}^{n} r(r+1)(r+2) = \frac{n}{4}(n+1)(n+2)(n+3)$

We can use these standard series to derive many other results such as

5. $\displaystyle\sum_{r=1}^{n} r^2 = \frac{n}{6}(n+1)(2n+1)$

6. $\displaystyle\sum_{r=1}^{n} r^3 = \frac{n^2}{4}(n+1)^2$

Mathematical Proof

Proof by Induction

To prove by mathematical induction, take the following steps:

1. By actual substitution, prove that the statement is true for $n = 1$, $n \in \mathbb{Z}^+$.
2. Assume that the statement is true for $n = k$, $k \in \mathbb{Z}^+$.
3. Prove that the statement is true for $n = (k + 1)$
   [This is the induction step]
4. Conclude that 'Therefore the statement is true $\forall n \in \mathbb{Z}^+$.

Example

Prove by induction that $\displaystyle\sum_{r=1}^{n} r(r+1) = \frac{1}{3}n(n+1)(n+2).$

Solution

Assume true for $n = k \Rightarrow \displaystyle\sum_{r=1}^{k} r(r+1) = \frac{1}{3}k(k+1)(k+2)$

Prove true for $k = 1$,

$\text{LHS} = \displaystyle\sum_{r=1}^{1} r(r+1) = 1(1+1) = 2$, $\text{RHS} = \frac{1}{3}(1)(2)(3) = 2$

Therefore, the statement is true for $n = 1$.

Prove that the statement is true for $n = k + 1$.

$$\sum_{r=1}^{k+1} r(r+1) = \sum_{r=1}^{k} r(r+1) + (k+1)(k+2)$$

$$= \frac{1}{3}k(k+1)(k+2) + (k+1)(k+2) = \frac{1}{3}(k+1)(k+2)(k+3)$$

$$\sum_{r=1}^{k+1} r(r+1) = = \frac{1}{3}(k+1)(k+2)(k+3)$$

Therefore, the statement is true for $n = k + 1$.

Proof by Contradiction

The principle of this proof is that if p is true then $\sim p$ is false and vice versa. To prove by contradiction,

1. State the negation $\sim p$ of the statement p.
2. Prove that $\sim p$ is false.
3. Conclude that 'the statement $\sim p$ is false $\Rightarrow$ the statement p is true.

Proof by Deduction

To prove by deduction, chronologically and systematically, use the implication of known and proven facts to arrive at the conclusion.

To prove that $p \Rightarrow q$ is true, start with p then deduce $p \Rightarrow r \Rightarrow s \Rightarrow q$, therefore $p \Rightarrow q$.

Example

Prove that if $ax^2 + bx + c = 0$, then $x = \frac{-b \pm \sqrt{b^2 - 4ac}}{2a}$.

This proof can be done by deduction. See topic 1 under "*Method of Completing the Square*" for this proof.

Proof by Counter Example

To prove by counter example, simply produce one case which shows that the statement is false.

TOPIC 11
PERMUTATIONS AND COMBINATIONS

Definitions

A permutation is an ordered *arrangement* of a number of objects. For instance, the permutations of (a, b, c) are

$(a, b, c), (a, c, b), (b, a, c), (b, c, a), (c, a, b), (c, b, a)$.

A combination is an ordered *selection* of a number of objects from among a given number of objects. For instance, the four different combinations of three letters from (a, b, c, d) are

$(a, b, c), (b, c, d), (c, d, a), (c, a, b)$.

The Factorial Notation

The product of all the positive integers to a given positive integer n is denoted by $n!$ read "n factorial".

$$n! = (1)(2)(3)(4)...(n-3)(n-2)(n-1)n$$

Conventionally, $1! = 1$ and $0! = 1$.

Permutations

1. The number of arrangements (permutations) of n different objects on a line is given by

    $$^{n}P_{n} = \frac{n!}{(n-n)!} = n!$$

2. The number of arrangements (permutations) of r objects chosen from among n objects is given by

    $$^{n}P_{r} = \frac{n!}{(n-r)!} = n(n-1)(n-2)(n-3)...(n-r+1)$$

3. The number of arrangements (permutations) of n different objects on a circle is given by $\dfrac{^{n}P_{n}}{n} = (n-1)!$

4. The number of arrangements (permutations) of n different objects on a ring is given by $\dfrac{^{n}P_{n}}{2n} = \dfrac{(n-1)!}{2}$

5. The number of arrangements (permutations) of n objects with p similar, q similar and r similar objects is given by

$$\frac{n!}{p!q!r!}.$$

6. The number of arrangements (permutations) of r objects chosen out of n different objects with repetition is given by n^r.

Combinations

The number of combination (selections) of r objects chosen out of n different objects is given by

$$^nC_r = \frac{n!}{(n-r)!r!} = {}^nC_{(n-r)} = \frac{^nP_r}{r!} = \frac{^nP_{(n-r)}}{(n-r)!}$$

Theorems on Permutations and Combinations

1. If one operation can be performed in m ways and another operation can be performed in n ways, then the two operations can be performed in mn ways.
2. If any of two operations can be performed in either m ways or n ways, then both operations can be performed in $m + n$ ways.

The Binomial Theorem

The expansion of the expression $(a+b)^n$ in powers of a and b is known as the binomial theorem.

Pascal's triangle

If n is small $n \in \mathbb{Z}^+$ the Pascal's triangle shown below, is a useful instrument for expanding $(a+b)^n$.

```
              1       1
          1       2       1
      1       3       3       1
  1       4       6       4       1
1     5      10      10      5     1
1   6    15      20      15    6    1
```

Generally,

$$(a+b)^n = a^n +^n C_1 a^{n-1}b +^n C_2 a^{n-2}b^2 +...+^n C_r a^{n-r}b^r +...+b^n$$

Remarks!

1. The general term is

$$^n C_r a^{n-r}b^r = \frac{n(n-1)(n-2)...(n-r+1)}{r!} a^{n-r}b^r .$$

2. The rth term is $u_r =^n C_{r-1} a^{n-r+1}b^{r-1}$.

3. The number of terms in the expansion is exactly $(n+1)$.

4. The expansion has no meaning unless when $n \in \mathbb{Z}^+$.

5. When $n \notin \mathbb{Z}^+$, the expansion has the following properties:

   (a) The expansion is valid only if $a = 1$ and $-1 < b < 1$.

   (b) Since $-1 < b < 1$ the series converges.

   (c) The expansion has an infinite number of terms.

   (d) Under the conditions in (a), (b) and (c) above the expansion can be rewritten as

   $$(1+x)^n = 1 + nx + \frac{n(n-1)}{2!}x^2 + \frac{n(n-1)(n-3)}{3!}x^3 +...$$

Example

Expand $(1 - 2x)^{-3}$ as a series of ascending powers of x up to and including the term in x^4 stating the range of values of x for which the expansion is valid.

Solution

$$(1 - 2x)^{-3} = 1 + (-3)(-2x) + \frac{(-3)(-4)(-2x)^2}{2!} +$$

$$\frac{(-3)(-4)(-5)(-2x)^3}{3!} + \frac{(-3)(-4)(-4)(-5)(-2x)^4}{4!} + ...$$

$$\Rightarrow (1 - 2x)^{-3} = 1 + 6x + 24x^2 + 80x^3 + 240x^4+...$$

Applications of Binomial Expansions in Approximations

The binomial expansion can be used to evaluate without tables or calculators and to a reasonable degree of accuracy the value of numerical expressions such as $(1.99)^{10}$. This can be done by expressing $(1.99)^{10}$ as $(2 - 0.01)^{10}$ and applying the binomial theorem.

Examples

1. Expand $(1 - 2x)^7$ In ascending powers of x up to and including the term in x^3 and hence find $(0.98)^7$ to 4 d. p.

2. Show that $\frac{1-x}{\sqrt{1+x}} = 1 - \frac{3}{2}x + \frac{7}{8}x^2$, if x is sufficiently small so x^3 and higher powers may be neglected.

3. When $(1 + bx)^n$ is expanded in ascending powers of x, the first few terms of the expansion are $1 - \frac{3}{5}x + \frac{27}{100}x^2$. Find the values of n and b.

4. Expand $\sqrt{\frac{1+x}{1-x}}$. By putting $x = \frac{1}{9}$, show that $\sqrt{5} \approx \frac{1630}{729}$.

Solutions

1. $(1 - 2x)^7 = 1 - 14x + 84x^2 - 280x^3$

   $(0.98)^7 = (1 - 0.02)^7 = (1 - 2(0.01))^7 \Rightarrow x = 0.01$

   $\Rightarrow (0.98)^7 = 1 - 14(0.01) + 81(0.01)^2 - 280(0.01)^3$

   $= 1 - 0.14 + 0.0081 - 0.000280 = 0.8681$ to 4 d. p.

2. $\frac{1-x}{\sqrt{1+x}} = (1 - x)(1 + x)^{-\frac{1}{2}}$,

   $$(1 + x)^{-\frac{1}{2}} = 1 + \left(-\frac{1}{2}\right)x + \left(-\frac{1}{2}\right)\left(-\frac{3}{2}\right)\frac{x^2}{2!} + \cdots$$

   $$\Rightarrow \frac{1 - x}{\sqrt{1 + x}} = (1 - x)\left(1 - \frac{1}{2}x + \frac{3}{8}x^2\right)$$

   $$= 1 - \frac{3}{2}x + \frac{7}{8}x^2$$

3. $(1 + bx)^n = 1 + nbx + n(n-1)(b^2)\left(\frac{x^2}{2}\right)$

Equating coefficients and arranging,

$$\frac{27}{50} = n^2 b^2 - nb^2 \text{ and } b = -\frac{3}{5n} \Rightarrow \frac{9}{25}(n-1) = \frac{27}{50}$$

$$\Rightarrow n = 2 \text{ and } b = -\frac{3}{10}$$

4. $\sqrt{\frac{1+x}{1-x}} = (1+x)^{\frac{1}{2}}(1-x)^{-\frac{1}{2}}$

$$(1+x)^{\frac{1}{2}} = 1 + \frac{1}{2}x + \frac{1}{2}\left(-\frac{1}{2}\right)\frac{x^2}{2!} + \frac{1}{2}\left(-\frac{1}{2}\right)\left(-\frac{3}{2}\right)\frac{x^3}{3!} + \cdots$$

$$(1-x)^{-\frac{1}{2}} = 1 + \left(-\frac{1}{2}\right)(-x) + \left(-\frac{1}{2}\right)\left(-\frac{3}{2}\right)\frac{x^2}{2!} +$$

$$\left(-\frac{1}{2}\right)\left(-\frac{3}{2}\right)\left(-\frac{5}{2}\right)\frac{(-x)^3}{3!}$$

$$(1+x)^{\frac{1}{2}}(1-x)^{-\frac{1}{2}} = \left(1 + \frac{1}{2}x - \frac{1}{8}x^2 + \frac{3}{16}x^3\right)$$

$$\left(1 + \frac{1}{2}x + \frac{3}{8}x^2 + \frac{5}{16}x^3\right)$$

$$\Rightarrow \sqrt{\frac{1+x}{1-x}} = 1 + x + \frac{1}{2}x^2 + \frac{7}{16}x^3 + \cdots$$

Putting $x = \frac{1}{9}$, $\frac{\sqrt{5}}{2} \approx \frac{814.9}{729} \Rightarrow \sqrt{5} \approx \frac{1630}{729}$ as required.

TOPIC 12
COORDINATE GEOMETRY

THE STRAIGHT LINE

Let $A(x_1, y_1)$ and $B(x_2, y_2)$ be two given points and $P(x, y)$ be any other point on a straight line, then;

1. The distance between A and B is given by

$$AB = \sqrt{(x_2 - x_1)^2 + (y_2 - y_1)^2}$$

2. The midpoint of the line segment AB is given by

$$M = \left(\frac{1}{2}(x_1 + x_2), \frac{1}{2}(y_1 + y_2) \right)$$

3. If θ is the angle between a line and the positive direction of the x-axis, the gradient of the line AB is given by

$$m = \frac{y_2 - y_1}{x_2 - x_1} = \tan\theta.$$

4. The gradient and two point form equation of the straight line is given by $(x_2 - x_1)(y - y_1) = (x - x_1)(y_2 - y_1)$

The gradient/intercept form of a straight line whose gradient is m and whose intercept with the y-axis is c is given by $y = mx + c$.

The general equation of a straight line is given by $ax + by + k = 0$, where a, b and k are constants.

Given two straight lines L_1 and L_2 with gradients m_1 and m_2 respectively then:

(a) L_1 is perpendicular to $L_2 \Rightarrow m_1 = m_2$.

(b) L_1 is parallel to $L_2 \Rightarrow m_1 m_2 = -1$.

(c) L_1 intersects $L_2 \Rightarrow$ their equations are both satisfied at the point of intersection. To find the point of intersection, solve simultaneously the equations of the straight lines L_1 and L_2.

(d)   If L_1 and L_2 intersect the angle θ between them is

given by $\tan\theta = \dfrac{m_1 - m_2}{1 + m_1 m_2}$

The perpendicular distance d of a point (x_1, y_1) from a straight line $ax + by + k = 0$ is given by

$$d = \left| \frac{ax_1 + by_1 + k}{\sqrt{a^2 + b^2}} \right|.$$

Reduction to Linear Form

It is often required to convert a non-linear relationship to a linear one, draw a straight line to represent it and then use it to find the values of certain constant parameters. To do this, express the relationship in the form $y = mx + c$ and compare the coefficients of the equation with those of $y = mx + c$.

Draw the graph and use it to find the gradient m and the intercept c with the y-axis, hence the values of the constant parameters in the non-linear relationship.

The following table shows some common non-linear relationships and their corresponding linear relationships.

	Non-linear Relationship	Linear Relationship	Comparison	
			m	c
1	$y = ax^n$	$\log y = n\log x + \log a$	n	$\log a$
2	$y = ab^x$	$\log y = x\log b + \log a$	$\log b$	$\log a$
3	$\dfrac{1}{x} + \dfrac{1}{y} = \dfrac{1}{a}$	$Y = -X + \dfrac{1}{a}$ $Y = \dfrac{1}{y}, X = \dfrac{1}{x}$	-1	$\dfrac{1}{a}$
4	$y = ax^2 + b$	$y = aX + b$ $X = x^2$	a	b

LOCI

Definition

A locus is the path of a moving point subject to some restrictions. The following are some common loci.

1. The locus of a point $P(x, y)$ whose distance r from a fixed point $O(x_0, y_0)$ is always constant is a circle.

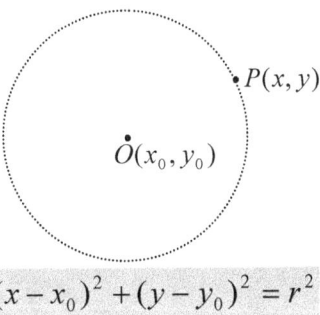

$$(x - x_0)^2 + (y - y_0)^2 = r^2$$

2. The locus of a point $P(x, y)$ whose distance d from two fixed points $A(x_1, y_1)$ and $B(x_2, y_2)$ is always equal is the mediator or perpendicular bisector of the line segment joining the two points.

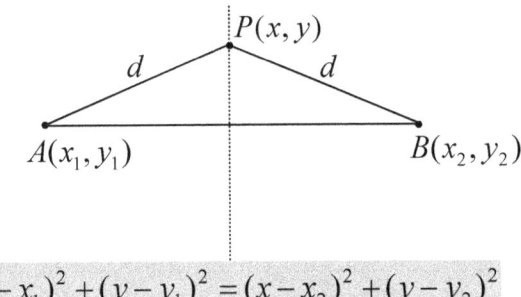

$$(x - x_1)^2 + (y - y_1)^2 = (x - x_2)^2 + (y - y_2)^2$$

3. The locus of a point, $P(x, y)$ whose distance d from two intersecting straight lines L_1 and L_2 is always equal, is the bisector of the angle between the two lines.

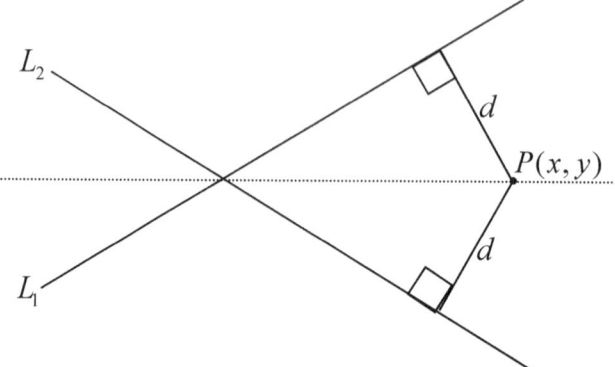

4. The locus of a point, $P(x, y)$ whose distance d from a fixed straight lines L_1 is always constant is a line L_2 parallel to the line L_1.

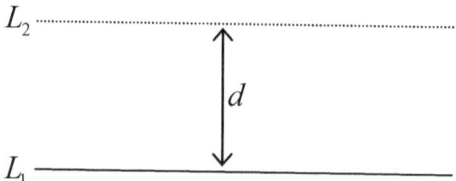

Circle Geometry
Some Useful Circle Theorems

1. The distance from any point $P(x_1, y_1)$ on the circle to the centre of the circle is equal to the radius of the circle. By implication, the distance of a tangent to a circle from the centre of the circle is equal to the radius of the circle. [This theorem is very useful in finding the equation of a tangent at a given point on a circle with given centre]

2. The angle subtended at the circumference by the diameter is $90°$ or simply stated; the angle in a semicircle is $90°$. [This theorem is very useful in finding the equation of a circle with two given points as the ends of a diameter].

3. The tangent to a circle is perpendicular to the radius of the circle at the point of contact. [This theorem is very useful in finding the equation of a tangent to a given circle at a given point]

4. The perpendicular bisector of a chord of a circle passes through the centre of the circle. [This theorem is very useful in finding the equation of a circle with three given points].

The Equation of a Circle

As earlier seen, the equation of a circle with centre $O(x_0, y_0)$ and radius r is $(x - x_0)^2 + (y - y_0)^2 = r^2$. The general form of the equation of a circle with centre $O(-g, -f)$ is
$x^2 + y^2 + 2gx + 2fy + c = 0$.
The radius of this circle is $r = \sqrt{g^2 + f^2 - c}$.
To find the centre and radius of a given circle, either complete the squares to have the equation in the form
$(x - x_0)^2 + (y - y_0)^2 = r^2$ or compare the given circle with the general equation of the circle.

Example
Find the centre and radius of the circle
$$x^2 + y^2 - 6x + 2y - 15 = 0$$

Solution
$(x - 3)^2 + (y + 1)^2 - 9 - 1 - 15 = 0$
$\Rightarrow (x - 3)^2 + (y + 1)^2 = 5^2$
Therefore, the centre of the circle is $(3, -1)$ and the radius is 5.
Alternatively, comparing $x^2 + y^2 - 6x + 2y - 15 = 0$ with the general equation of the circle, $2g = -6 \Rightarrow g = -3$,
$2f = 2 \Rightarrow f = 1$ and $c = -15$. Therefore, the centre is
$(3, -1)$ and the radius is $r = \sqrt{(-3)^2 + (1)^2 - (-15)} = 5$.

Relationship between a Straight Line and a Circle (Application of Nature of Roots of Quadratic Equations)

A circle $x^2 + y^2 + 2gx + 2fy + c = 0$ may relate to a straight line $y = mx + k$ in the following three different ways.

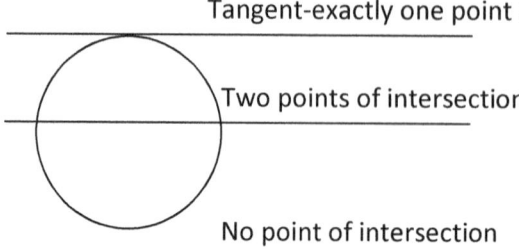

Tangent-exactly one point of contact

Two points of intersection

No point of intersection

To determine the relationship, substitute the equation of the line into that of the circle to obtain a quadratic equation of the form $(1 + m^2)x^2 + 2(fm + g + km)x + k^2 + c = 0$.

1. The straight line intersects with the circle at two distinct points hence, the equation of the circle and the line, are simultaneously satisfied. In this case, the resulting quadratic equation has two real and distinct roots. The implication is that the discriminant is greater than zero.
2. The line touches the circle at exactly one point. In other words, the line is a tangent to the circle. In this case, the resulting quadratic equation has exactly one real root. The implication is that the discriminant is equal to zero.
3. The line neither intersects the circle nor is a tangent to the circle. In this case, the resulting quadratic equation has no real roots. The implication is that the discriminant is less than zero.

Example
Show that the circle $x^2 + y^2 - 8x + 6y + 5 = 0$ intersects with the line $y = -2x + 5$ at two distinct points and find their points of intersection.

Solution
$x^2 + (-2x + 5)^2 - 8x + 6(-2x + 5) + 5 = 0$
$\Rightarrow x^2 - 8x + 12 = 0$ and $\Delta = (-8)^2 - 4(1)(12) = 16 > 0$.
Since $\Delta > 0$, they intersect at two distinct points.
$x^2 - 8x + 12 = 0 \Rightarrow x = 2$ or $x = 6$.
These two values of x give $y = 1$ or $y = -7$ respectively.
Therefore, their points of intersection are $(2,1)$ and $(6, -7)$.

Length of a Tangent from an External Point to a Circle

The length $d = PT$ of a tangent from an external point $P(x_1, y_1)$ to a circle with centre $C(x_0, y_0)$ is given by

$$d^2 = (x_1 - x_0)^2 + (y_1 - y_0)^2 - r^2$$

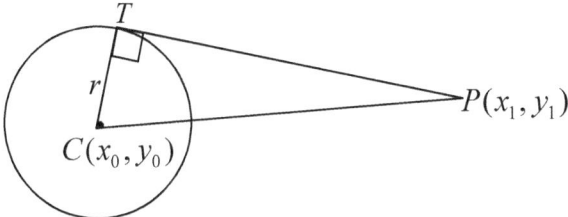

Equation of Circle on $A(x_1, y_1)$ and $B(x_2, y_2)$ as diameter

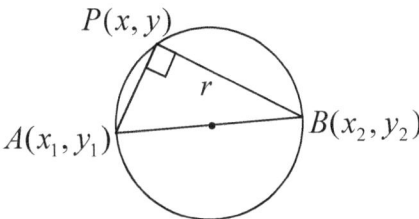

The products of the gradients of the lines PA and PB is -1, since they are perpendicular to each other.

$$\Rightarrow \left(\frac{y - y_1}{x - x_1} \right)\left(\frac{y - y_2}{x - x_2} \right) = -1$$

Rearranging gives

$$\Rightarrow x^2 + y^2 - (x_1 + x_2)x - (y_1 + y_2)y + x_1 x_2 + y_1 y_2 = 0$$

Orthogonal and Touching Circles

Given two circles S_1 and S_2 with centres C_1 and C_2 and radii r_1 and r_2 respectively, then

1. S_1 and S_2 touch externally if $r_1 + r_2 = C_1 C_2$ (fig (i)) below.

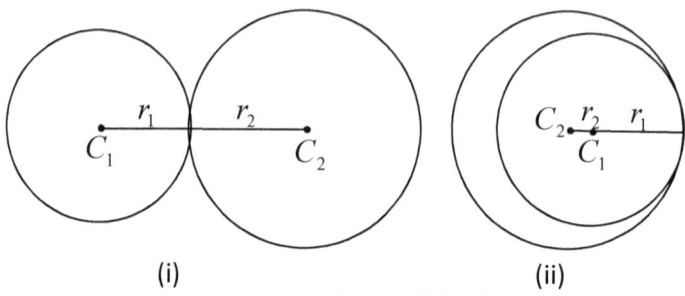

(i)   (ii)

2. S_1 and S_2 touch internally if $|r_1 - r_2| = C_1C_2$ (fig (ii)) above.

3. S_1 and S_2 are orthogonal (at right angles to each other) if
$r_1^2 + r_2^2 = (C_1C_2)^2$.

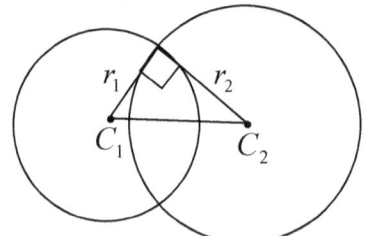

Some Special Circle Related Equations

1. The equation of the common chord or radical axis of two
   circles $S_1 = 0$ and $S_2 = 0$ is given by $S_1 - S_2 = 0$.

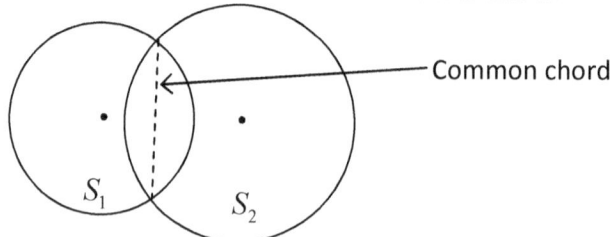

Common chord

2. The equation of the circle passing through the points of
   intersection of two circles $S_1 = 0$ and $S_2 = 0$ (fig (i))
   below is given by $S_1 + kS_2 = 0$, where k is a constant.

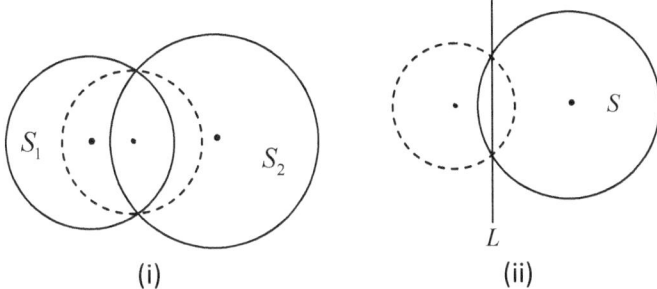

(i)   (ii)

3. The equation of a circle passing through the points of intersection of the circle $S = 0$ and the straight line $L = 0$ (fig (i)) above is given by $S + kL = 0$, where k is a constant.

4. The equation of a circle that touches the x-axis (fig (i)) below is given by $(x - r)^2 + (y - f)^2 = r^2$.

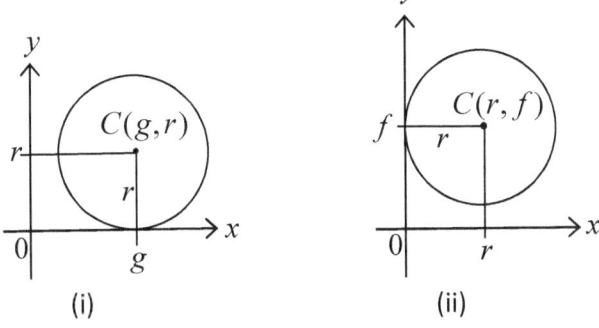

(i)   (ii)

5. The equation of a circle that touches the y-axis (fig (ii)) above is given by $(x - g)^2 + (y - r)^2 = r^2$.

6. The equation of a circle that touches the x and y-axes (fig (i)) below is given by $(x - r)^2 + (y - r)^2 = r^2$.

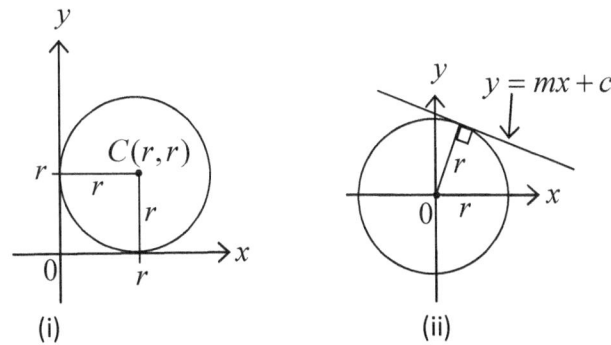

(i)   (ii)

7. The condition that the line $y = mx + c$ is a tangent to the circle $x^2 + y^2 = r^2$ (fig (ii)) above is that $c = \pm r\sqrt{1 + m^2}$. This condition is obtained by finding the distance of the centre $(0,0)$ from the line $y = mx + c$.

8. The equation of a circle that touches the straight line $ax + by + c = 0$ and whose centre is $C(x_1, y_1)$ is given by

$$(x - x_1)^2 + (y - y_1)^2 = r^2, \text{ where } r = \frac{ax_1 + by_1 + c}{\sqrt{a^2 + b^2}}.$$

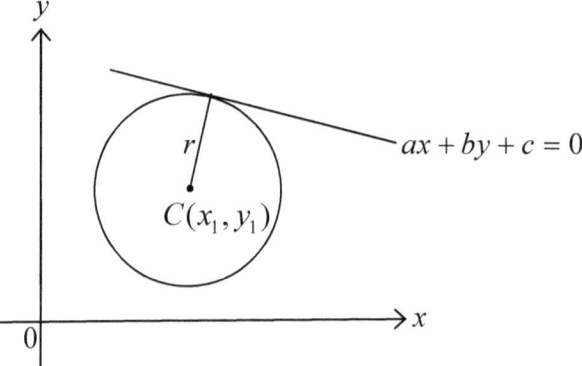

Appreciate that r is the distance of the centre $C(x_1, y_1)$ from the circle and hence the line $ax + by + c = 0$.

9. The equation of a tangent to a circle with centre $C(x_1, y_1)$ at the point $P(x_2, y_2)$ is given by $y = mx + c$, where

$$m = -\frac{x_2 - x_1}{y_2 - y_1} \text{ and } c = y_2 - mx_2.$$

Appreciate that gradient of radius $\times m = -1$ and that the tangent passes through P.

10. The equation of a tangent from the origin to a circle with centre $C(x_1, y_1)$ and radius r is given by $y = mx$, where m is obtained from the equation $r = \frac{mx_1 - y_1}{\sqrt{m^2 + 1}}$.

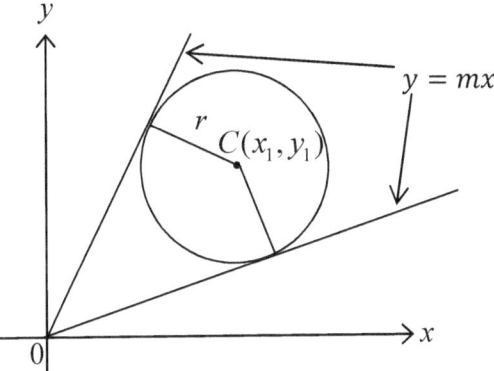

The essential fact to remember in 7, 8 and 10 is the formula for finding the distance of a point from a line and that the distance of the centre from the circle and hence from the tangent is equal to the radius r of the circle.

TOPIC 13
COMPLEX NUMBERS

A complex number is of the form $a + bi$ where a and b are real numbers and $i = \sqrt{-1}$ is called an imaginary unit. $i^2 = -1$. The set of complex numbers is denoted by $\mathbb{C}$.

Thus $\mathbb{C} = \{x : x = a + bi, a, b \in \mathbb{R}\}$.

In the complex number $z = a + bi$, a is called the real part and b is called the imaginary part of the complex number. Thus if $z = a + bi$, Real $Z = \mathrm{Re}Z = a$ and Imaginary $Z = \mathrm{Im}Z = b$.

Algebra of Complex Numbers

$\forall a, b, c, d \in \mathbb{R},$

1. $(a + bi) + (c + di) = (a + c) + (b + d)i.$

2. $(a + bi) - (c + di) = (a - c) + (b - d)i.$

3. $(a + bi)(c + di) = (ac - bd) + (ad + bc)i.$

4. $\dfrac{(a+bi)}{(c+di)} = \dfrac{(a+bi)}{(c+di)} \dfrac{(c-di)}{(c-di)} = \dfrac{ac+bd}{c^2+d^2} + \left(\dfrac{bc-ad}{c^2+d^2}\right)i,$
    where $c^2 + d^2 \neq 0.$

The complex conjugate of $z = a + bi$ denoted by z^* or $\bar{z}$ is $a - bi$ and vice versa. Thus $z^* = \bar{z} = a - bi.$
$(a + bi)(a - bi) = a^2 + b^2$

Properties of Complex Conjugates

1. $z_1{}^* + z_2{}^* = (z_1 + z_2)^*$

2. $z_1{}^* - z_2{}^* = (z_1 - z_2)^*$

3. $z_1{}^* z_2{}^* = (z_1 z_2)^*$

4. $(z^*)^* = z, \quad \dfrac{1}{z^*} = \left(\dfrac{1}{z}\right)^*$

5. zz^* and $z + z^*$ are both real numbers.

6. $z_1 - z^* = 2\mathrm{Im}\, z, \quad z_1 + z^* = 2\mathrm{Re}\, z$

Complex Roots of Polynomial Equations

The complex roots of any polynomial equation occur in conjugate pairs. Thus if one root is $z = a + bi$, another root must be $z = a - bi$.

Equality of Complex Numbers

Two complex numbers $z_1 = a + bi$ and $z_2 = c + di$ are equal if $a = c$ and $b = d$.

The Square Roots of a Complex Number

Let $\sqrt{a + bi} = x + yi \Rightarrow a + bi = x^2 - y^2 + 2xyi$.

$$\Rightarrow a = x^2 - y^2 \dots\dots\dots\dots\dots\dots\dots\text{①}$$

$$\text{And } b = 2xy \dots\dots\dots\dots\dots\dots\dots\dots\text{②}$$

We can find the values of x and y by solving ① and ② simultaneously.

Example

Evaluate $\sqrt{8 - 6i}$.

Solution

Let $\sqrt{8 - 6i} = x + yi \Rightarrow 8 - 6i = x^2 - y^2 + 2xyi$.

$$\Rightarrow 8 = x^2 - y^2 \dots\dots\dots\dots\dots\dots\dots\text{①}$$

$$\text{And } -3 = xy \dots\dots\dots\dots\dots\dots\dots\dots\text{②}$$

From ②, $9 = x^2 y^2 \Rightarrow y^2 = \frac{9}{x^2} \dots\dots\dots\text{③}$

Substitute ③ in ①:

$$8 = x^2 - \frac{9}{x^2} \Rightarrow x^4 - 8x^2 - 9 = 0.$$

$$(x^2 - 9)(x^2 + 1) = 0 \Rightarrow x = \pm 3 \text{ or } x = \pm i.$$

Using ②, when $x = 3, y = -1$ and when $x = -3, y = 1$.

85

$\therefore \sqrt{8 - 6i} = 3 - i$ or $\sqrt{8 - 6i} = -3 + i$

Substituting $x = \pm i$ is needless because the roots are in cyclic order and lead to the same result.

Argand Diagrams

We can represent the complex number $z = x + yi$ in a two dimensional plane just as the ordered pair (x, y) by placing the real part on the x-axis and the imaginary part on the y-axis. This explains why the x-axis is often referred to as the real axis and the y-axis as the imaginary axis.

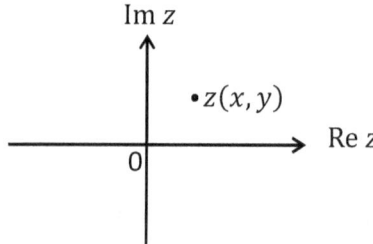

Modulus and Argument of a Complex Number

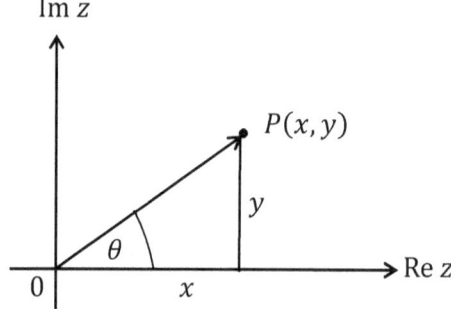

The modulus of the complex number $z = x + yi$ denoted by $|z| = |x + yi|$ is the length of the line segment OP where P is the point (x, y) and O is the origin.

The modulus of the complex number $z = x + yi$ is defined as $|z| = |x + yi| = \sqrt{x^2 + y^2}$.

The principal argument of $z = x + yi$ denoted by $\arg z$ is the angle which OP makes with the real axis and $-\frac{\pi}{2} \leq \arg z \leq \frac{\pi}{2}$. The general solution of $\arg z$ denoted by $\text{Arg } z$ is given by

$$\text{Arg } z = \text{Arg}(x + yi) = \tan^{-1}\left(\frac{y}{x}\right).$$

Note! In dealing with arguments, it is advisable to always draw an Argand diagram to see clearly the quadrant in which the angles lie.

Example

Find the modulus and argument of each of the following:

(a) $1 + i\sqrt{3}$ (b) $-1 + i\sqrt{3}$ (c) $-1 - i\sqrt{3}$ (d) $1 - i\sqrt{3}$

Solution

For all the cases, the modulus $r = 2$.
The arguments are as follows:

(a) $\frac{\pi}{3}$ (b) $\frac{2\pi}{3}$ (c) $\frac{4\pi}{3}$ (d) $\frac{5\pi}{3}$

Modulus-Argument Form or Polar Form of a Complex Number

We can express any complex number $z = x + yi$ in the form $z = r(\cos\theta + i\sin\theta)$, where r is the modulus and θ is the argument of $z = x + yi$. Thus

$$z = x + yi \Leftrightarrow z = r(\cos\theta + i\sin\theta)$$

$r = \sqrt{x^2 + y^2}$ and $\theta = \tan^{-1}\left(\frac{y}{x}\right)$, where $-\frac{\pi}{2} \leq \theta \leq \frac{\pi}{2}$ is the principal argument.

The conjugate of $z = r(\cos\theta + i\sin\theta)$ is

$z = r(\cos\theta - i\sin\theta) = r\{\cos(-\theta) + i\sin(-\theta)\}$.

Example

Express $z = 2 + 2i$ in the form $z = r(\cos\theta + i\sin\theta)$.

Solution

$$r = \sqrt{x^2 + y^2} = \sqrt{2^2 + 2^2} = 2\sqrt{2}$$

$$\theta = \tan^{-1}\left(\frac{y}{x}\right) = \tan^{-1}\left(\frac{2}{2}\right) = \tan^{-1}(1) = \frac{\pi}{4}$$

$$\Rightarrow z = 2\sqrt{2}\left(\cos\frac{\pi}{4} + i\sin\frac{\pi}{4}\right)$$

Multiplication and Division in Polar Form

If $z_1 = r_1(\cos\theta_1 + i\sin\theta_1)$ and
$z_2 = r_2(\cos\theta_2 + i\sin\theta_2)$. Then,

$$|z_1 z_2| = r_1 r_2 \text{ and } \arg(z_1 z_2) = \arg z_1 + \arg z_2$$

$$\left|\frac{z_1}{z_2}\right| = \frac{r_1}{r_2} \text{ and } \arg\left(\frac{z_1}{z_2}\right) = \arg z_1 - \arg z_2$$

$$\Rightarrow z_1 z_2 = r_1 r_2\{\cos(\theta_1 + \theta_2) + i\sin(\theta_1 + \theta_2)\}$$

$$\frac{z_1}{z_2} = \frac{r_1}{r_2}\{\cos(\theta_1 - \theta_2) + i\sin(\theta_1 - \theta_2)\}$$

De Moivre's Theorem

De Moivre's Theorem states that if $z = \cos\theta + i\sin\theta$, then
$z^n = (\cos\theta + i\sin\theta)^n = (\cos n\theta + i\sin n\theta)$.

We can use this theorem to express $\sin n\theta$, $\cos n\theta$ and $\tan n\theta$
in terms of powers of $\sin\theta$, $\cos\theta$ and $\tan\theta$. For instance,

$$\cos 5\theta = \text{Re}(\cos 5\theta + i\sin 5\theta) = \text{Re}(\cos\theta + i\sin\theta)^5$$

$$(c + is)^5 = c^5 + 5c^4 is + 10c^3(is)^2 + 10c^2(is)^3 + 5c(is)^4 + (is)^5$$

$$\Rightarrow \cos 5\theta = \cos^5\theta - 10\cos^3\theta\sin^2\theta + 5\cos\theta\sin^4\theta$$

$$\sin 5\theta = \text{Im}(\cos 5\theta + i\sin 5\theta) = \text{Im}(\cos\theta + i\sin\theta)^5$$

$$\Rightarrow \sin 5\theta = 5\cos^4\theta\sin\theta - 10\cos^2\theta\sin^3\theta + \sin^5\theta$$

$$\tan 5\theta = \frac{\sin 5\theta}{\cos 5\theta} = \frac{5\cos^4\theta\sin\theta - 10\cos^2\theta\sin^3\theta + \sin^5\theta}{\cos^5\theta - 10\cos^3\theta\sin^2\theta + 5\cos\theta\sin^4\theta}$$

Dividing each term by $\cos^5\theta$, we have

$$\tan 5\theta = \frac{5\tan\theta - 10\tan^3\theta + \tan^5\theta}{1 - 10\tan^2\theta + 5\tan^4\theta}.$$

TOPIC 14
VECTORS

Definition

A **vector quantity** is a quantity which has both magnitude and direction. Examples of vector quantities are displacement, velocity, force etc.

A **scalar quantity** is a quantity which has only magnitude. Examples of scalar quantities are mass, temperature, distance etc.

Representation of Vectors

A vector can be represented by a directed line segment AB.

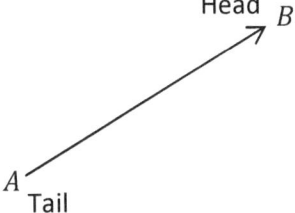

AB represents a displacement from the point A to the point B and the arrow shows the direction of the vector.

Unit Vector

A unit vector is a vector whose magnitude is one or unity.

A unit vector in the direction of the vector **a** is denoted by $\hat{\mathbf{a}} = \frac{\mathbf{a}}{|\mathbf{a}|}$.

Components of a Vector

A vector $\mathbf{r} = \mu\mathbf{a} + \lambda\mathbf{b}$ has two components $\mu\mathbf{a}$ in the direction of **a** and $\lambda\mathbf{b}$ in the direction of **b**.

Base Vectors

The vectors **i** and **j** are orthogonal unit base vectors in two dimensions. This means that **i** and **j** are unit vectors perpendicular to each other and every other vector in two dimensions can be represented in terms of **i** and **j**.
Thus $\mathbf{r} = x\mathbf{i} + y\mathbf{j}$.

The vectors **i, j** and **k** are orthogonal unit base vectors in three dimensions. This means that **i, j** and **k** are unit vectors perpendicular to each other and every other vector in three dimensions can be represented in terms of **i, j** and **k**.
Thus $\mathbf{r} = x\mathbf{i} + y\mathbf{j} + z\mathbf{k}$.

Column Vectors

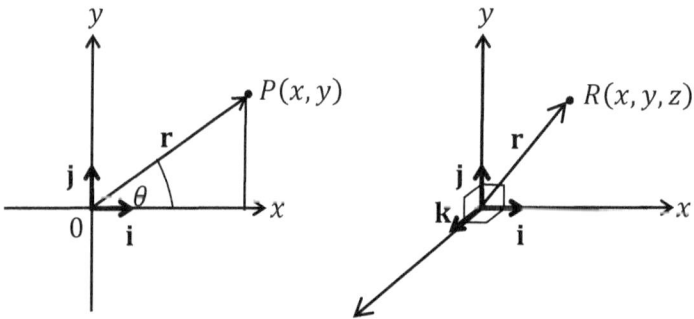

The vectors $\mathbf{r} = x\mathbf{i} + y\mathbf{j}$ and $\mathbf{r} = x\mathbf{i} + y\mathbf{j} + z\mathbf{k}$ can be

represented in column vector form as $\mathbf{r} = \begin{pmatrix} x \\ y \end{pmatrix}$ and $\mathbf{r} = \begin{pmatrix} x \\ y \\ z \end{pmatrix}$

respectively, where

x represents the number of units in the **i** direction,

y represents the number of units in the **j** direction and

z represents the number of units in the **k** direction.

Magnitude and Direction of a Vector

The magnitude of a vector **OP** represents the length of the line segment **OP** and is given by $OP = \sqrt{x^2 + y^2}$ in two dimensions or $OP = \sqrt{x^2 + y^2 + z^2}$ in three dimensions.

The direction of **OP** is the angle which **OP** makes with the positive direction of the x-axis.

In two dimensions, the direction of $\mathbf{OP} = \begin{pmatrix} x \\ y \end{pmatrix} = x\mathbf{i} + y\mathbf{j}$ is given by $\theta = \tan^{-1}\left(\frac{y}{x}\right)$.

Position Vectors

The position vector of a point A denoted by **OA** is a vector which is tied to the origin. If A is the point $(1, -2, 3)$ then the position vector of A is $\mathbf{OA} = \mathbf{i} - 2\mathbf{j} + 3\mathbf{k}$. **AB** is the position vector of the point B relative to the point A and is given by

$$\mathbf{AB} = \mathbf{AO} + \mathbf{OB} = \mathbf{OB} - \mathbf{OA}.$$

Vector Algebra

1. **Equality of Vectors**
    Two vectors are equal if they have the same magnitude and direction.

    $$\mathbf{a} = \mathbf{b} \Rightarrow |\mathbf{a}| = |\mathbf{b}| \text{ and direction of } \mathbf{a} = \text{direction of } \mathbf{b}$$

2. **Addition and Subtraction of Vectors**
    Let $\mathbf{a} = x_1\mathbf{i} + y_1\mathbf{j} + z_1\mathbf{k}$ and $\mathbf{b} = x_2\mathbf{i} + y_2\mathbf{j} + z_2\mathbf{k}$.
    Then $\mathbf{a} + \mathbf{b} = (x_1 + x_2)\mathbf{i} + (y_1 + y_2)\mathbf{j} + (z_1 + z_2)\mathbf{k}$
    $$\mathbf{a} - \mathbf{b} = (x_1 - x_2)\mathbf{i} + (y_1 - y_2)\mathbf{j} + (z_1 - z_2)\mathbf{k}$$
    As column vectors, $\mathbf{a} + \mathbf{b} = \begin{pmatrix} x_1 \\ y_1 \\ z_1 \end{pmatrix} + \begin{pmatrix} x_2 \\ y_2 \\ z_2 \end{pmatrix} = \begin{pmatrix} x_1 + x_2 \\ y_1 + y_2 \\ z_1 + z_2 \end{pmatrix}$

$$a - b = \begin{pmatrix} x_1 \\ y_1 \\ z_1 \end{pmatrix} - \begin{pmatrix} x_2 \\ y_2 \\ z_2 \end{pmatrix} = \begin{pmatrix} x_1 - x_2 \\ y_1 - y_2 \\ z_1 - z_2 \end{pmatrix}$$

3. Multiplication of a Vector by a Scalar

If **a** and **b** are parallel vectors then, **a** can be represented as a scalar multiple of **b**. i.e. **a** ∥ **b** ⟹ **a** = λ**b**, λ ∈ ℝ .

4. Addition and Subtraction of displacement Vectors
(a) Addition

Diagrammatically, vectors are added head to tail.

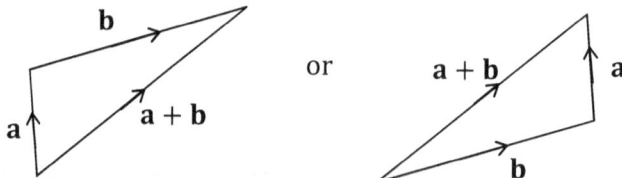

The above diagrams show the two ways in which the vectors **a** and **b** may be added. This demonstrates that vector addition is **commutative**. i.e. **a** + **b** = **b** + **a**.
If the two figures are combined as below, we can see that the sum of the two vectors is represented by the diagonal of the parallelogram. This is called the parallelogram law of vector addition.

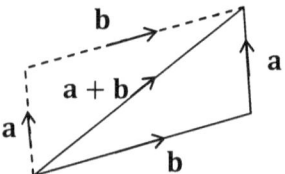

Given more than two vectors, the vectors can still be arranged head to tail and then added as illustrated below for the addition of the vectors **a**, **b** and **c**.This is known as the polygon law of vector addition and shows that vector addition obeys the **associative law**.

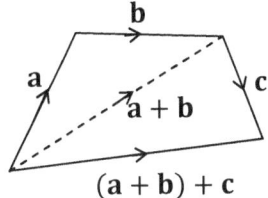 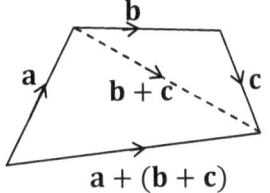

5. **Additive inverse of a Vector**

   The additive inverse – **a** of a vector **a** is a vector with the same magnitude as **a** but opposite direction.

   The sum of a vector and its additive inverse is a vector whose magnitude is zero. A vector whose magnitude is zero is called a **zero** or **null vector** denoted by **0**. Thus

   $$\mathbf{a} + (-\mathbf{a}) = -\mathbf{a} + \mathbf{a} = \mathbf{0}$$

(b) **Subtraction**

   Subtracting a vector **b** is the same as adding its additive inverse – **b**. Thus $\mathbf{a} - \mathbf{b} = \mathbf{a} + (-\mathbf{b})$. The implication of the above statement is that diagrammatically, vectors are subtracted tail to tail, or head to head.

   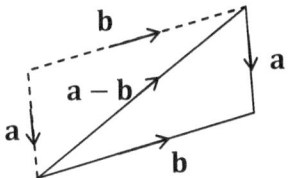

   Therefore, in the parallelogram of vectors, the sum of the vectors is obtained by measuring the diagonal which connects the tail of one vector to the head of the other while the difference of the vectors is obtained by measuring the diagonal which connects the tails of the vectors or the heads of the vectors.

6. The section or Ratio Theorem

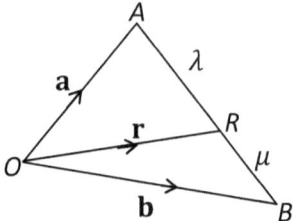

Given that the point R whose position vector is $\mathbf{r}$ divides the line segment joining the points A and B whose position vectors are $\mathbf{a}$ and $\mathbf{b}$ in the ratio $\lambda:\mu$, then,

$$\mathbf{r} = \frac{\mu\mathbf{a} + \lambda\mathbf{b}}{\lambda + \mu}$$

If $\lambda + \mu = 1$, $\mathbf{r} = \mu\mathbf{a} + \lambda\mathbf{b}$.

If R is the midpoint of $\mathbf{AB}$, then $\lambda = \mu \Rightarrow \mathbf{r} = \frac{1}{2}(\mathbf{a} + \mathbf{b})$.

7. Dot or Scalar Product

The dot or scalar product of two vectors $\mathbf{a}$ and $\mathbf{b}$ denoted by $\mathbf{a} \cdot \mathbf{b}$ and read $\mathbf{a}$ dot $\mathbf{b}$ is given by $\mathbf{a} \cdot \mathbf{b} = |\mathbf{a}||\mathbf{b}| \cos \theta$, where θ is the angle between the two vectors.
From the definition it follows that,

$\mathbf{a} \perp \mathbf{b} \Leftrightarrow \mathbf{a} \cdot \mathbf{b} = 0$ $\mathbf{a} \parallel \mathbf{b} \Leftrightarrow |\mathbf{a}||\mathbf{b}| = 0$

From the above it is clear that,
$\mathbf{i} \cdot \mathbf{i} = \mathbf{j} \cdot \mathbf{j} = \mathbf{k} \cdot \mathbf{k} = 1$ and
$\mathbf{i} \cdot \mathbf{j} = \mathbf{j} \cdot \mathbf{k} = \mathbf{k} \cdot \mathbf{i} = \mathbf{j} \cdot \mathbf{i} = \mathbf{k} \cdot \mathbf{j} = \mathbf{i} \cdot \mathbf{k} = 0$

Therefore $\boldsymbol{a} \cdot \boldsymbol{b}$ is a scalar.
$(x_1\boldsymbol{i} + y_1\boldsymbol{j} + z_1\boldsymbol{k}) \cdot (x_2\boldsymbol{i} + y_2\boldsymbol{j} + z_2\boldsymbol{k}) = x_1x_2 + y_1y_2 + z_1z_2$

Properties of the Dot Product
1. Commutativity: $\boldsymbol{a} \cdot \boldsymbol{b} = \boldsymbol{b} \cdot \boldsymbol{a}$
2. Scalar multiplication: $\lambda(\boldsymbol{a} \cdot \boldsymbol{b}) = (\lambda\boldsymbol{a}) \cdot \boldsymbol{b} = \boldsymbol{a} \cdot (\lambda\boldsymbol{b})$
3. Distributivity over addition: $\boldsymbol{a} \cdot (\boldsymbol{b} + \boldsymbol{c}) = \boldsymbol{a} \cdot \boldsymbol{b} + \boldsymbol{a} \cdot \boldsymbol{c}$

The Vector (or Cross) Product

The vector product $\mathbf{a} \times \mathbf{b}$ of two vectors $\mathbf{a} = x_1\mathbf{i} + y_1\mathbf{j} + z_1\mathbf{k}$ and $\mathbf{b} = x_2\mathbf{i} + y_2\mathbf{j} + z_2\mathbf{k}$ is defined as the vector whose magnitude is equal to $ab\sin\theta$, where θ is the angle between $\mathbf{a}$ and $\mathbf{b}$. The vector product $\mathbf{a} \times \mathbf{b}$ is given by

$$\mathbf{a} \times \mathbf{b} = \begin{vmatrix} \mathbf{i} & \mathbf{j} & \mathbf{k} \\ x_1 & y_1 & z_1 \\ x_2 & y_2 & z_2 \end{vmatrix} = \begin{vmatrix} y_1 & z_1 \\ y_2 & z_2 \end{vmatrix}\mathbf{i} - \begin{vmatrix} x_1 & z_1 \\ x_2 & z_2 \end{vmatrix}\mathbf{j} + \begin{vmatrix} x_1 & y_1 \\ x_2 & y_2 \end{vmatrix}\mathbf{k}$$

$$\mathbf{a} \times \mathbf{b} = (y_1 z_2 - y_2 z_1)\mathbf{i} - (x_1 z_2 - x_2 z_1)\mathbf{j} + (x_1 y_2 - x_2 y_1)\mathbf{k}$$

$$|\mathbf{a} \times \mathbf{b}| = ab\sin\theta$$

The vector $\mathbf{a} \times \mathbf{b}$ is perpendicular to the plane containing the vectors $\mathbf{a}$ and $\mathbf{b}$.

Vector Equation of a Straight Line

1. The vector equation of a straight line L which is parallel to a vector $\mathbf{b}$ and passes through a point whose position vector is $\mathbf{a}$ is given by $\mathbf{r} = \mathbf{a} + \lambda\mathbf{b}$.

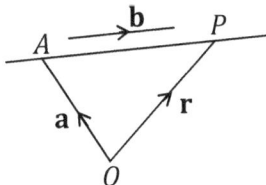

2. The vector equation of a straight line L which passes through two fixed points whose position vectors are $\mathbf{a}$ and $\mathbf{b}$ is given by $\mathbf{r} = \mathbf{a} + \lambda(\mathbf{b} - \mathbf{a})$.

Note that the vector equation of a straight line is not unique because the point $\mathbf{a}$ is just one of the number of points on the line.

Cartesian Equation of a Straight Line

The Cartesian equation of a straight line which passes through the point (x_1, y_1, z_1) and is parallel to the vector $a\mathbf{i} + b\mathbf{j} + c\mathbf{k}$ is given by

$$\frac{x - x_1}{a} = \frac{y - y_1}{b} = \frac{z - z_1}{c} (= \lambda)$$

The Cartesian equation of a straight line which passes through the points $A(x_1, y_1, z_1)$ and $B(x_2, y_2, z_2)$ is given by

$$\frac{x - x_1}{x_2 - x_1} = \frac{y - y_1}{y_2 - y_1} = \frac{z - z_1}{z_2 - z_1} (= \lambda)$$

Parametric Equations of a Straight Line

The parametric equations of a straight line which passes through the point (x_1, y_1, z_1) and is parallel to the vector $a\mathbf{i} + b\mathbf{j} + c\mathbf{k}$ are given by

$$x = x_1 + \lambda a, \quad y = y_1 + \lambda b, \quad z = z_1 + \lambda b.$$

Direction Ratios and Direction Cosines

The direction ratios of a straight line which is parallel to the vector $a\mathbf{i} + b\mathbf{j} + c\mathbf{k}$ are given by $a:b:c$. The values a, b and c which occur in all the three forms of the equations of a straight line determine the direction of the line.

The direction cosines of a line which is in the direction of the vector $a\mathbf{i} + b\mathbf{j} + c\mathbf{k}$ are given by

$$l = \cos \alpha = \frac{a}{\sqrt{a^2 + b^2 + c^2}}$$

$$m = \cos \beta = \frac{b}{\sqrt{a^2 + b^2 + c^2}}$$

$$n = \cos \gamma = \frac{c}{\sqrt{a^2 + b^2 + c^2}}$$

Where α, β and γ are the angles which the line makes with the positive directions of the x, y and z-axes.

Pair of Lines

Let $L_1: \mathbf{r} = \mathbf{a_1} + \lambda \mathbf{b_1}$ and $L_2: \mathbf{r} = \mathbf{a_2} + \lambda \mathbf{b_2}$, then

1. If L_1 and L_2 represent the same straight line, then $\mathbf{b_1} = \mathbf{b_2}$.
2. If L_1 and L_2 are parallel, $\mathbf{b_1} = k\mathbf{b_2}$, where k is a scalar.
3. If L_1 and L_2 intersect then there exist unique values of λ and μ such that, $\mathbf{a_1} + \lambda \mathbf{b_1} = \mathbf{a_2} + \lambda \mathbf{b_2}$.
4. If L_1 and L_2 are skew [skew lines are lines which are not parallel and do not intersect] there exist no unique value of λ and μ such that $\mathbf{a_1} + \lambda \mathbf{b_1} = \mathbf{a_2} + \lambda \mathbf{b_2}$. For instance, $\mathbf{r} = \mathbf{i} + 3\mathbf{k} + \lambda(2\mathbf{i} + \mathbf{j} + \mathbf{k})$ and $\mathbf{r} = 2\mathbf{i} - \mathbf{j} + \mathbf{k} + \mu(\mathbf{i} - 2\mathbf{j})$ are skew since the values of λ and μ obtained by equating the coefficients of $\mathbf{i}$ and $\mathbf{j}$ do not satisfy the coefficients of $\mathbf{z}$.

5. The angle θ between L_1 and L_2 is given by $\cos \theta = \frac{\mathbf{b_1} \cdot \mathbf{b_2}}{|\mathbf{b_1}||\mathbf{b_2}|}$

   The angle θ between $\frac{x-x_1}{a_1} = \frac{y-y_1}{b_1} = \frac{z-z_1}{c_1}$ and $\frac{x-x_2}{a_2} = \frac{y-y_2}{b_2} = \frac{z-z_2}{c_2}$ is given by

$$\cos \theta = \frac{a_1 a_2 + b_1 b_2 + c_1 c_2}{\left(\sqrt{a_1^2 + b_1^2 + c_1^2}\right)\left(\sqrt{a_2^2 + b_2^2 + c_2^2}\right)}$$

   The angle θ between two lines L_1 and L_2 whose direction cosines are l_1, m_1, n_1 and l_2, m_2, n_2 is given by
   $\cos \theta = l_1 l_2, + m_1 m_2 + n_1 n_2$

Perpendicular Distance of a Point to a Given Straight Line

The perpendicular distance d from a point P with position vector $\mathbf{p}$ to a line L whose equation is $\mathbf{r} = \mathbf{a} + \lambda \mathbf{b}$ can be obtained by finding λ from the equation $(\mathbf{a} + \lambda_1 \mathbf{b} - \mathbf{p}) \cdot \mathbf{b} = 0$ and substituting in the equation $d = |\mathbf{a} + \lambda_1 \mathbf{b} - \mathbf{p}|$.

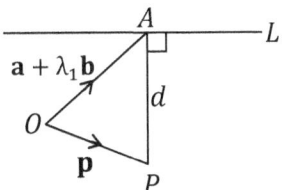

The Shortest Distance between Two Skew Lines

The shortest distance d between two skew lines $\mathbf{r} = \mathbf{a}_1 + \lambda\mathbf{b}_1$ and $\mathbf{r} = \mathbf{a}_2 + \mu\mathbf{b}_2$ is given by $d = |(\mathbf{a}_1 - \mathbf{a}_2) \cdot (\mathbf{b}_1 \times \mathbf{b}_2)|$

Planes

A plane is a flat surface.

Equations of a Plane

Given that $\mathbf{r}$ is the position vector of any point on a plane and $\mathbf{n} = a\mathbf{i} + b\mathbf{j} + c\mathbf{k}$ is a vector perpendicular to the plane.

1. If $\mathbf{a}_1$ is the position vector of a fixed point on the plane, $\mathbf{b}$ and $\mathbf{c}$ are two non-parallel vectors on the plane and λ and μ are scalars Then the parametric equation of the plane is given by $\mathbf{r} = \mathbf{a}_1 + \lambda\mathbf{b} + \mu\mathbf{c}$.

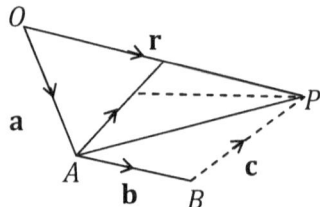

The only difference in the equations of two parallel planes $\pi_1: \mathbf{r} = \mathbf{a}_1 + \lambda\mathbf{b} + \mu\mathbf{c}$ and $\pi_1: \mathbf{r} = \mathbf{a}_2 + \lambda\mathbf{b} + \mu\mathbf{c}$ is their distance from the origin.

2. If d the distance of the origin from the plane, the scalar product form of the equation of the plane is given by $\mathbf{r} \cdot \mathbf{n} = d$.

3. The vector equation of a plane which contains the point with position vector $\mathbf{a}_1$ and is perpendicular to the unit vector $\hat{\mathbf{n}}$ is given by $\mathbf{r} \cdot \hat{\mathbf{n}} = \mathbf{a} \cdot \hat{\mathbf{n}}$.

4. The Cartesian equation of the plane is given by $ax + by + cz = d$.

5. The equation of the plane which passes through three given point $\mathbf{a}_1, \mathbf{a}_2, \mathbf{a}_3$ is given by
$$\{(\mathbf{a}_3 - \mathbf{a}_2) \times (\mathbf{a}_3 - \mathbf{a}_1)\} \cdot \mathbf{r} = \{(\mathbf{a}_3 - \mathbf{a}_2) \times (\mathbf{a}_3 - \mathbf{a}_1)\} \cdot \mathbf{a}_1$$

Note that $\mathbf{n} = (\mathbf{a}_3 - \mathbf{a}_2) \times (\mathbf{a}_3 - \mathbf{a}_1)$ is a vector perpendicular to the plane.

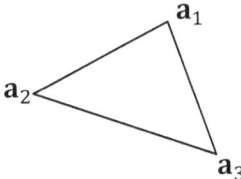

6. To find the equation of the plane which passes through a given point $\mathbf{a}_1$ and which contains a given line $\mathbf{r} = \mathbf{a}_2 + \lambda\mathbf{b}$, substitute $\lambda = 1$ in the equation $\mathbf{r} = \mathbf{a}_2 + \lambda\mathbf{b}$ to find a third point $\mathbf{a}_3$ and use the formula in 5.

7. The equation of the plane which passes through a given point $\mathbf{a}_1$ and is parallel to a given plane $\mathbf{r} \cdot \mathbf{n} = d$ is given by $(\mathbf{r} - \mathbf{a}_1) \cdot \mathbf{n} = 0$.

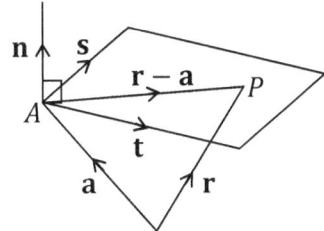

8. The equation of the plane which passes through the intersection of the planes $\mathbf{r} \cdot \hat{\mathbf{n}}_1 = d_1$ and $\mathbf{r} \cdot \hat{\mathbf{n}}_2 = d_2$ is given by $\mathbf{r} \cdot (\hat{\mathbf{n}}_1 - k\hat{\mathbf{n}}_2) = d_1 - kd_2$.

Intersection of Lines and Planes

1. The equation of the line which passes through a given point $\mathbf{a}_1$ and is perpendicular to a given plane $\mathbf{r} \cdot \mathbf{n} = d$ is given by $\mathbf{r} = \mathbf{a}_1 + \lambda\mathbf{n}$.

2. To find the line of intersection of two planes $\mathbf{r} \cdot \mathbf{n}_1 = d_1$ and $\mathbf{r} \cdot \mathbf{n}_2 = d_2$, first find a point $\mathbf{a}_1$ common to both planes. This is done by taking for convenience one of x, y or z to be zero and solving the Cartesian equations of the two planes simultaneously. Since the point $\mathbf{a}_1$ is on the line of intersection of two planes. The line of intersection of the two planes $\mathbf{r} \cdot \mathbf{n}_1 = d_1$ and $\mathbf{r} \cdot \mathbf{n}_2 = d_2$ is then given by $\mathbf{r} = \mathbf{a}_1 + \lambda(\mathbf{n}_1 \times \mathbf{n}_2)$.

99

Example

Find the vector equation of the line in which the two planes $2x - 5y + 3z = 12$ and $3x + 4y - 3z = 6$ meet.

Solution

Solving the planes simultaneously:

When $x = 0$, $y = -18$ and $z = -26 \Longrightarrow a_1 = -18\mathbf{j} - 26\mathbf{k}$

$(3\mathbf{i} + 4\mathbf{j} - 3\mathbf{k}) \times (2\mathbf{i} - 5\mathbf{j} + 3\mathbf{k}) = 3\mathbf{i} + 5\mathbf{j} + 23\mathbf{k}.$

Therefore, the equation of the required line is
$\mathbf{r} = -18\mathbf{j} - 26\mathbf{k} + \lambda(3\mathbf{i} + 5\mathbf{j} + 23\mathbf{k}).$

3. The point of intersection of the line $\mathbf{r} = \mathbf{a} + \lambda\mathbf{b}$ and the plane $\mathbf{r} \cdot \mathbf{n} = d$ is obtained by solving the equation $(\mathbf{a} + \lambda\mathbf{b}) \cdot \mathbf{n} = d$ and $\mathbf{r} = \mathbf{a} + \lambda\mathbf{b}$ simultaneously. This is done by finding the value of λ from the equation $(\mathbf{a} + \lambda\mathbf{b}) \cdot \mathbf{n} = d$ and substituting in $\mathbf{r} = \mathbf{a} + \lambda\mathbf{b}$.

4. The angle θ between the line $\mathbf{r} = \mathbf{a} + \lambda\mathbf{b}$ and the plane $\mathbf{r} \cdot \hat{\mathbf{n}} = d$ is given by

$$\sin\theta = \frac{\mathbf{b} \cdot \hat{\mathbf{n}}}{|\mathbf{b}|}$$

5. The angle θ between two planes is equal to the angle between their two normal vectors and is given by

$$\cos\theta = \hat{\mathbf{n}}_1 \cdot \hat{\mathbf{n}}_2 \text{ or } \cos\theta = \frac{\mathbf{n}_1 \cdot \mathbf{n}_2}{|\mathbf{n}||\mathbf{n}|}.$$

Where $\hat{\mathbf{n}}_1$ and $\hat{\mathbf{n}}_2$ are the unit normal vectors of the two given planes, and θ is the angle between the two planes.

Perpendicular Distance from a Point to a Plane

The perpendicular distance from the point (x_1, y_1, z_1) to the plane $ax + by + cz + d$ is given by

$$d = \frac{ax_1 + by_1 + cz_1 + d}{\sqrt{a^2 + b^2 + c^2}}$$

TOPIC 15
BINARY RELATIONS

Definition and Notation of Binary Relations

A binary relation is a rule that assigns an element $x \in A$ to another element $y \in A$ (relations in a set) or a rule that assigns an element $x \in A$ to another element $y \in B$ (relations from one set to another). Some examples of relations are "is less than" e.g. 2<5, "is a subset of" e.g. $\{1,2\} \subset \{1,2,3,4\}$, "is a brother of" e.g. Eric is a brother of Jane.

Ordered Pairs

An ordered pair denoted by (a, b) is a pair of elements such that a is the first element and b is the second element.

Note that $(a, b) \neq (b, a)$.

Cartesian product of Two Sets

The Cartesian product $A \times B$ of two sets A and B is the set of all ordered pairs (a, b) such that $a \in A$ and $b \in B$ Thus

$$A \times B = \{(a,b): a \in A, b \in B\}$$

If $A = B = \mathbb{R}$ the set of real numbers then, $\mathbb{R} \times \mathbb{R} = \mathbb{R}^2$ is the entire Cartesian plane.

Intuitively, a relation is a statement involving two objects a and b. Therefore, a relation is either true or false depending on the values of a and b.

If "a relates b" we write $a \Re b$ or $(a, b) \in \Re$.

If "a does not relate b" we write $a \not\Re b$ or $(a, b) \notin \Re$.

Properties of Relations in a Set

Let $\Re$ be a relation in a set A then:

1. $\Re$ is **reflexive** if $a\Re a, \forall a \in A$. In other words, every element is related to itself.

2. $\Re$ is **symmetric** if $a\Re b \Rightarrow b\Re a, \forall a \in A$.

3. $\Re$ is **anti-symmetric** if $\forall a, b \in A, a\Re b$ and $b\Re a \Rightarrow a = b$.

4. $\Re$ is **transitive** if $\forall a, b, c \in A, a\Re b$ and $b\Re c \Rightarrow a\Re c$.

Equivalence Relations

An equivalence relation in a set A is a relation, which is reflexive, symmetric and transitive. As a mental aid, think of reflexive, symmetric and transitive as the letters RST.

Equivalence Classes

If $\Re$ is an equivalence relation on a non-empty set A, then the equivalence class of an element $a \in A$ denoted by $[a]$ is the set of all elements which belong to A and are related to a. i.e. the equivalence class of a is generated by a and is defined by $[a] = \{b \in A: b\Re a\}$. E.g. The relation $\Re$ defined on the set $\mathbb{Z}$ by $a\Re b \Rightarrow a - b = 2k$, $k\in\mathbb{Z}$ is an equivalence relation.

The class $[5] = \{x \in \mathbb{Z}: 5 - x = 2k, k \in \mathbb{Z}\}$. Thus,

$[5] = \{\cdots, -3, -1, 1, 3, 5, 7, \cdots\}$. Also $[5] = [x], \forall x \in [5]$.

Every x in $[5]$ is said to be a representative of $[5]$.

Partitions

A partition is a division of a set into subsets so that each of its elements is in exactly one subset.
For instance, let $A = \{a, b, c, d, e, f, g\}$, $B_1 = \{a, b, e\}$, $B_2 = \{c, g\}$, $B_3 = \{d, f\}$. Then $\{B_1, B_2, B_3\}$ is a partition of A.

The fundamental property of partitions is that if $\{A_1, A_2, \cdots, A_n\}$ is a partition of A, then

(i) $A = A_1 \cup A_2 \cup \cdots \cup A_n$. i.e. A is the union of all the sets.

(ii)   Members of A are **pair-wise disjoint**. In other words, the intersection of any two of the sets is an empty set.

$\{A_1, A_2, \cdots, A_n\}$ is a partition of $A \Rightarrow A_i \cap A_j = \emptyset, i \neq j$

and $i, j \in \mathbb{N}$.

For instance, if $A = \{2,4,6,8,10,12,14\}$ then clearly

$A_1 = \{2,6\}$, $A_2 = \{4\}$, $A_3 = \{8,10,14\}$ and $A_4 = \{12\}$, are subsets of A. Therefore $F = \{A_1, A_2, A_3, A_4\}$ forms a partition of A.

Venn diagram Illustration

Suppose $F = \{A_1, A_2, A_3, A_4\}$, then the relationship between the elements of F can be expressed on a Venn diagram as follows.

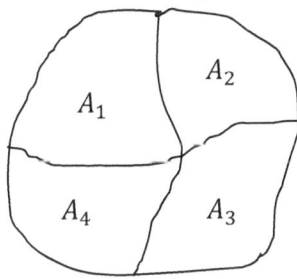

Fundamental Theorem on Equivalence Relations

An equivalence relation $\Re$ on a set A partitions the set A by putting those elements which are related to each other in the same equivalence class. In other words if $\Re$ is an equivalence relation on the set A and for every $a \in A$, if $[a] = \{x \in A: a\Re x\}$, then the family of sets $\{[a], a \in A\}$, is a partition of A.

The Quotient Set

If $\Re$ is an equivalence relation on the set A, then the quotient set denoted by $a/\Re$ is the set of all equivalence classes.

Example

The relation $\Re$ is defined on the set $A = \{1,2,3,4,5\}$ A as $a\Re b \Rightarrow a + b$ is even$\forall a, b \in A$. Verify that $\Re$ is an equivalence relation partitioned into two classes.

Solution

Since $1\Re 1, 1\Re 3, 1\Re 5, 3\Re 3, 3\Re 5$ and $5\Re 5$, $A_1 = [1] = \{1,3,5\}$ is one of the equivalence classes.

Similarly $2\Re 2, 2\Re 4$ and $4\Re 4$ so $A_2 = [2] = \{2,45\}$ is the other equivalence class. $\Rightarrow A = \{A_1, A_2\}$ is the partition of A

And $A/\Re = \{[1], [2]\} = \{A_1, A_2\}$

Order Relations

(i) Partial Order

A relation $\Re$ on a non-empty set A is a partial order if $\Re$ is reflexive, anti-symmetric and transitive. The mnemonic RAT may be a mental aid. Examples of partial order relations are:

(a) The relation "is a subset of" on the power set $P(A)$.

(b) The relation "is less than or equal to" on $\mathbb{R}$ the set of real numbers.

Inclusion Diagrams

A partial order relation on a finite set can be displayed using an inclusion diagram. For instance, the relation / defined on the set $A = \{1,2,3,4,6,12\}$ by $a/b \Rightarrow a$ divides b is a partial order and the inclusion diagram is as follows.

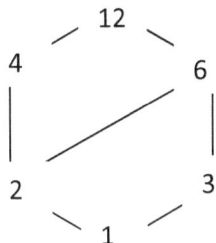

(ii) Total Order

A relation $\Re$ on a non-empty set A is a total order (or weak order) if $\Re$ is transitive, anti-symmetric and either $a\Re b$ or $b\Re a$ (or both). In such a case A and B are said to be comparable. For instance, let $A = \{\{1\}, \{1,2\}, \{1,2,3\}\}$. Then, the set inclusion relation $\subseteq$ defined on the set A is a total order (or weak order) since $X\Re Y$ means $\forall X, Y \in A, \ X \subseteq Y$.

(iii) Strict Order

Let $\Re$ be a weak order relation on the set A. Then, the corresponding strict order relation S is given by xSy means precisely $x\Re y$ and $x \neq y$. Hence if $\Re$ is the relation $\leq$ then S will be the relation $<$.

Therefore, a strict order relation S on a set A is a relation which is transitive and aSb or $bSa, a \neq b, \forall \ a, b \in A$.

TOPIC 16
MAPPINGS AND FUNCTIONS

MAPPINGS

A mapping is a correspondence between two sets, the domain and the codomain. Members of the codomain to which are mapped to by members of the domain form a set called the range or image set.

Notation

We denote a mapping from a set X to a set Y as $X \longrightarrow Y$.

If $x \in X$ and $y \in Y$ and $x\Re y$, we write $x \mapsto y$. Notice the difference between the two arrows $\longrightarrow$ and $\mapsto$.

We denote a mapping by a letter symbol such as f and represent it by $f: X \longrightarrow Y$, $f(x) = y$ or $f: x \mapsto y$ where y is the image of x under f. We can use any other letter such as g, h, k etc in place of f.

Representation of Mappings

We can represent a mapping $f: X \longrightarrow Y$ between two sets $X = \{a, b, c\}$ and $Y = \{1,2,3\}$ arrow or pappy diagrams as follows.

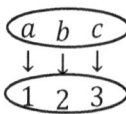 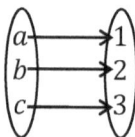

Classification of Mappings

A **one-to-one mapping** is a mapping in which one and only one element in the domain A, is mapped to one and only one element in the codomain B as in the diagrams above.

106

Let f be a mapping which assigns to each country its capital city then f is a one-to-one mapping. The domain of f is the list of countries and the codomain is the list of capital cities.

A **many-to-one mapping** is a mapping in which two or more elements in the domain map to the same image in the codomain.

A mapping $f: \mathbb{R} \rightarrow \mathbb{R}$ defined by $f: x \longmapsto x^2$ is a many-to-one mapping since for instance the image of -3 is 9 and that of 3 is also 9.

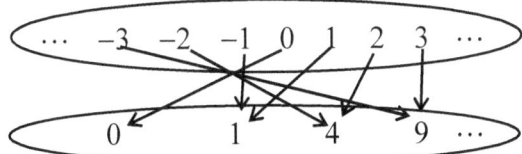

A **many-to-many mapping** is a mapping in which two or more elements in the domain map to two or more images in the codomain.

A relation defined from the set P of parents to the set C of children is a many-to-many mapping since every child has two parents (mother and father) and a father or mother may have more than one child.

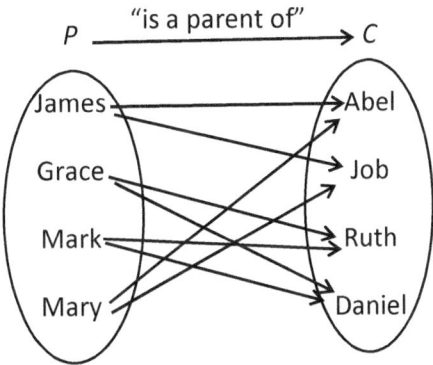

A **one-to-many mapping** is a mapping in which a single element in the domain maps to more than one image in the codomain.

A relation defined from the set M of mothers to the set C of children is a one-to-many mapping since every child has one mother and a mother may have more than one child.

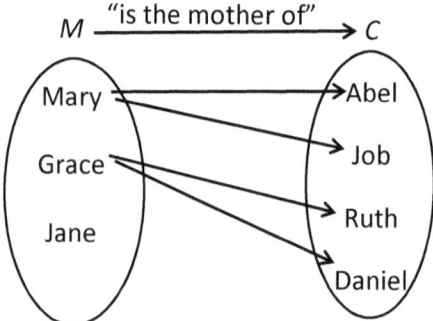

An "**onto**" mapping is a mapping in which all the elements of the codomain are "used up".

A relation defined from the set M of mothers to the set C of children is an on-to mapping since every child has one mother.

An "**into**" mapping is a mapping in which all the elements of the codomain are not "used up".

A relation defined from the set C of children to the set M of mothers is an "into" mapping since it is possible for a mother to be childless.

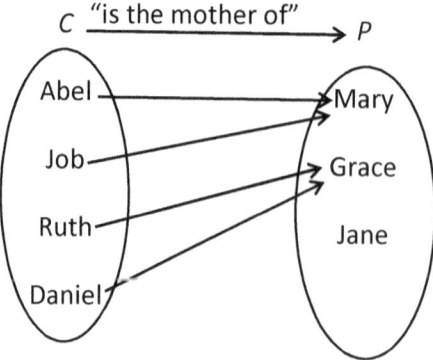

FUNCTIONS

A function is a one-to-one or a many-to-one mapping. An example of a function is the mapping $f: \mathbb{R} \to \mathbb{R}$ defined by $f: x \mapsto x^2$.

Equality of Functions

Given that $f: A \to B$ and $g: P \to Q$, then the functions f and g are equal $\Leftrightarrow A = P, B = Q$ and $f(a) = g(a)$, $\forall a \in A$.

Surjective Function

A surjection or surjective function is a function in which every element in the codomain is an image of an element in the domain.

$$\text{Image set of } f = \text{codomain of } f \Leftrightarrow \text{surjection}$$

A surjection is an "onto" mapping. For instance, given that $f: \mathbb{R}^+ \to \mathbb{R}^+$ where $f(x) = x^2$, then f is a surjection on $\mathbb{R}^+$.

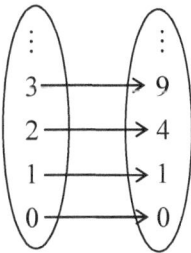

For a surjection, the cardinality of the codomain B is always less than or equal to that of the domain A.

$$n(B) \leq n(A)$$

Injective Function

An injection or injective function is a function which is such that every element $x \in A$ has a unique image $y \in B$. In other words an injection is a one-to-one mapping of two sets such that each element of the domain corresponds to only one element of the codomain. For instance;

Let $A = \{1,2,3,4\}$ and $B = \{1,2,3,4,5,6\}$. The function $f: A \longrightarrow B$, defined by $f: x \longmapsto x + 1$, shown below is injective.

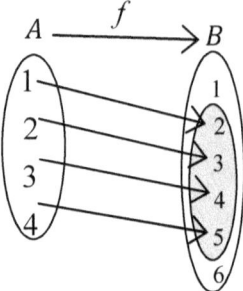

For an injection, the cardinality of the domain A is always less than or equal to that of the codomain B.

$$n(A) \leq n(B)$$

An injection is an into-mapping.

Bijective Function

A bijection or bijective function is a function that is both injective (into) and surjective (onto). Therefore a bijection is a one-one mapping for which all the elements in the codomain are mapped to. In other words, a bijection is a mapping between two sets in which every element in each set corresponds to only one element of the other set. For instance;

The function $f: A \longrightarrow B, f: x \longmapsto x + 1$ from $A = \{1,2,3,4\}$ to $B = \{2,3,4,5\}$ shown in below is injective and surjective. Hence the function is bijective.

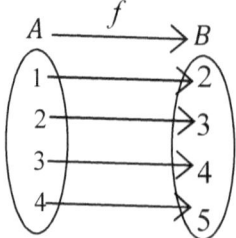

To prove that a function f is a bijection it suffices to show that

(a) Image of f = codomain of f

(b) $f(a) = f(b) \Rightarrow a = b$.

Composite Functions

A **composite function** $f \circ g$ or fg, is a function which is made up of two or more simpler functions f and g. A composite function such as $f \circ g \circ h$ by convention is operated from right to left. The figure below shows $h = f \circ g(x) = 2x + 5$, the result of composing set A using the function $g: x \mapsto 2x$ followed by $f: x \mapsto x + 5$.

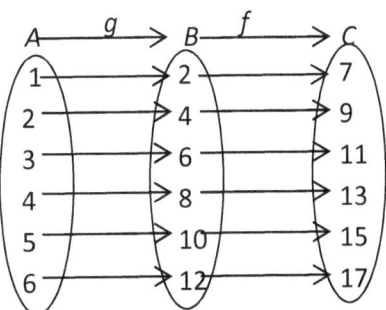

Identity Function

An identity function I is a function defined on a non-empty set A as $I: A \longrightarrow A$ and $I(a) = a, \forall a \in A$. In other words the identity function always maps every element to itself.

An identity function is a bijection.

Example

Given that $f: x \mapsto 2x + 1$ and $g: x \mapsto x + 3x^2$, express $f \circ g$ in the form $f \circ g: x \mapsto \cdots$.

Solution

$$f \circ g(x) = f(x + 3x^2) = 2(x + 3x^2) + 1$$

$$\Rightarrow f \circ g: x \mapsto = 6x^2 + 2x + 1$$

Example

The functions f and g are defined on $\mathbb{R}$, the set of real numbers, by $f: x \longmapsto x + 4, g: x \longmapsto \frac{1}{x+2}, x \neq 2$. Evaluate $f \circ g(2)$

Solution

$$f \circ g(2) = f\left(\frac{1}{2+2}\right) = f\left(\frac{1}{4}\right) = \frac{1}{4} + 4 = \frac{17}{4}$$

Alternatively,

$$f \circ g(x) = f\left(\frac{1}{x+2}\right) = \frac{1}{x+2} + 4 = \frac{4x+9}{x+2}$$

$$f \circ g(2) = \frac{4(2)+9}{2+2} = \frac{17}{4}$$

Flow Charts

Functions are often analyzed using flow charts.

Example

Draw a flow chart to represent the function $f: x \longmapsto \frac{9-7x}{6}$.

Solution

$$x \rightarrow \boxed{\times(-7)} \xrightarrow{-7x} \boxed{+9} \xrightarrow{9-7x} \boxed{\div 6} \rightarrow \frac{9-7x}{6}$$

Notice that in flow charts the order of operation is of utmost importance. For instance in the above function, it is imperative to first multiply x by -7 before adding 9 to the result.

Inverse Function

The inverse f^{-1} of a bijective function f is a function that performs the reverse process of what the function f does. For instance, if $g: x \mapsto x - 2$ then $g^{-1}: x \mapsto x + 2$. A function f which has an inverse is said to be invertible.

Note that:
(i) Only bijections are invertible.
(ii) If the domain and the codomain are equal, then
$ff^{-1} = f^{-1}f = I$, where I is the identity function.

Finding the Inverse of a function

Suppose that $f: x \mapsto f(x)$. To find f^{-1}, interchange the roles of x and y and solve for the new y. In other words substitute x for y and y for x and solve for the new y to have $f^{-1}(x)$.

Example

Given that $f: x \mapsto \dfrac{2x-7}{5}$, find the inverse of f.

Solution

Let $\dfrac{2y-7}{5} = x \Rightarrow y = \dfrac{5x+7}{2} \Rightarrow f^{-1}: x \mapsto \dfrac{5x+7}{2}$.

Alternatively; to find the inverse of f draw a flow chart of the function and used it to draw an inverse flow chart as shown below.

$$x \to \boxed{\times 2} \xrightarrow{2x} \boxed{-7} \xrightarrow{2x-7} \boxed{\div 5} \to \dfrac{2x-7}{5}$$

$$\dfrac{5x+7}{2} \leftarrow \boxed{\div 2} \xleftarrow{5x+7} \boxed{+7} \xleftarrow{5x} \boxed{\times 5} \leftarrow x$$

$$\Rightarrow f^{-1}: x \mapsto \dfrac{5x+7}{2}$$

The flow chart method has a disadvantage that it is very difficult to use when the function is very complicated.

Restriction of a Function

Consider the function $f: x \longmapsto x^2$. The domain of definition of this function is $\mathbb{R}$ to $\mathbb{R}$. Clearly, every element in the codomain except 0 is the image of two elements in the domain as shown in the arrow diagram (i) below.

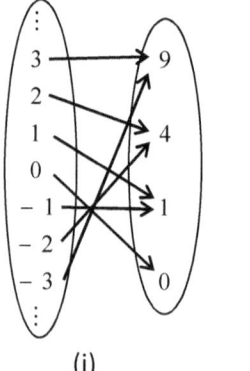   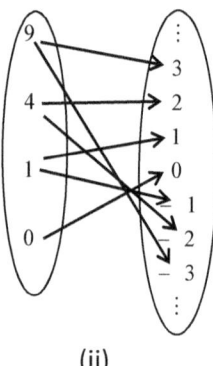

(i)   (ii)

The arrow diagram representing the inverse relation will be as shown in (ii). Therefore, though the relation is a function, the inverse relation is not a function. Hence the function $f: \mathbb{R} \longrightarrow \mathbb{R}$; $f: x \longmapsto x^2$ has no inverse. Suppose the function is redefined as $f: \mathbb{R}^+ \longrightarrow \mathbb{R}^+$; $f: x \longmapsto x^2$ the arrow diagram for the relation and the inverse relation will be as shown in (i) and (ii) below.

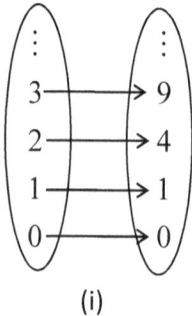   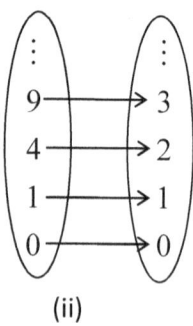

(i)   (ii)

Clearly both the relation and the inverse relation are functions. The new domain $\mathbb{R}^+ = [0, +\infty)$ of the function f is called the **restricted domain** of the function. A restricted domain is therefore the use of a domain for a function that is smaller than the function's domain of definition. Restricted domains are commonly used to specify a one-to-one section of a function. A function with a restricted domain is called a **restricted function**.

Another example is the function $f : x \mapsto \dfrac{1}{x}$. The set $\mathbb{R}$ of real numbers cannot be the domain of this function, since there is no value for $f(0)$. Therefore the domain of this function is the restricted domain $\mathbb{R} - \{0\}$

Relation between a function and its Inverse

The range of a function is the domain of its inverse f^{-1} and vice versa. i.e. $\mathrm{Im} f = Df^{-1}$ and $Df = \mathrm{Im}\, f^{-1}$.

Note that:

(i)   If f^{-1} exist, then it must be unique.

(ii)   If f is invertible so is f^{-1} and $(f^{-1})^{-1} = f$.

(iii)   If f and g have inverses, so does $g \circ f$ and $(g \circ f)^{-1} = f^{-1} \circ g^{-1}$.

Example

Given that $f: \mathbb{R} - (-1\} \longrightarrow \mathbb{R} - \{2\}$ where $f(x) = \dfrac{2x+1}{x+1}$, then $f^{-1}(x) = -\dfrac{x-1}{x-2}$.

$\mathrm{Im} f = \mathbb{R} - \{2\} = Df$ and $\mathrm{Im} f^{-1} = \mathbb{R} - \{-1\} = Df$.

The graph of f^{-1} is the reflection of the graph of f in the line $y = x$.

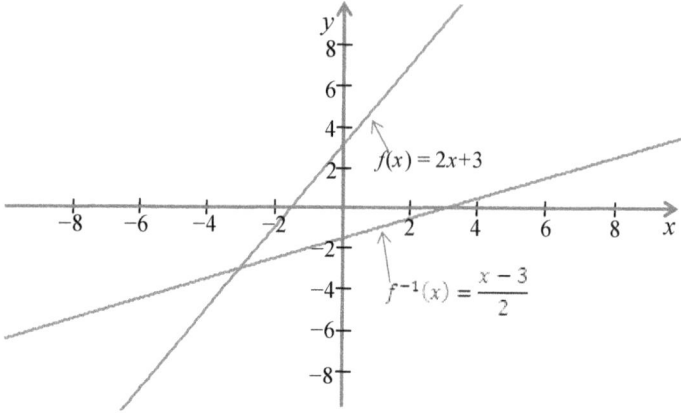

Types of Functions

1. **Even Function**

   An even function $f(x)$ is a function defined for positive and negative values of x for which $f(-x) = f(x)$ i.e. $f(a) = b \implies f(-a) = b$.

   Graphs of even functions are symmetrical about the y-axis (the line $x = 0$). The following graphs illustrate some examples of even functions.

$$f(x) = x^2$$

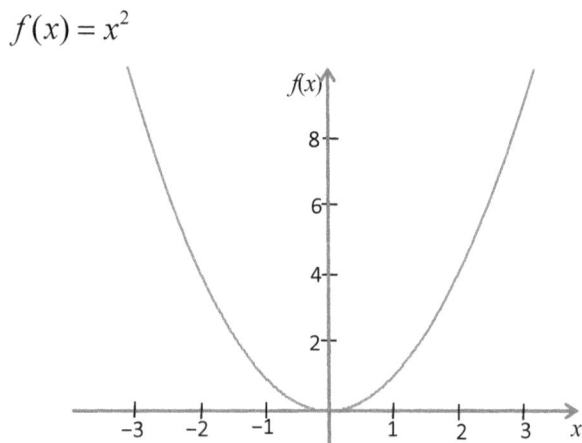

$f(x) = \cos x$

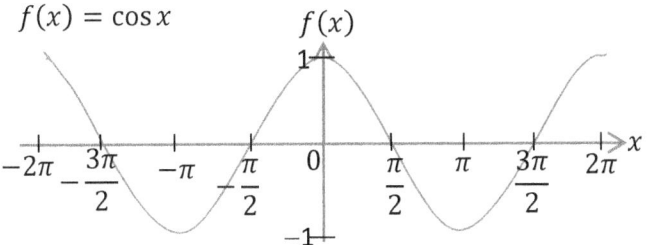

2. Odd Function

An odd function $f(x)$ is a function defined for positive and negative values of x for which $f(-x) = -f(x)$ i.e. $f(a) = b \Rightarrow f(-a) = -b$.

Graphs of odd functions are symmetrical about the line $f(x) = -x$. The following graphs illustrate some examples of even functions.

$f(x) = \sin x$

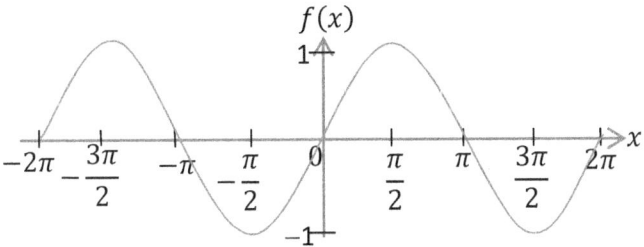

$$f(x) = \frac{1}{x}$$

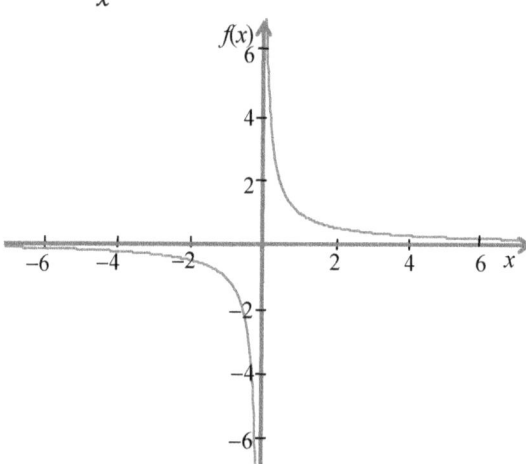

Some functions are neither odd nor even.

E.g. $f(x) = x^2 + \sin x$.

The following are graphs of some functions which are neither even nor odd.

$$f(x) = x^3 - x + 1$$

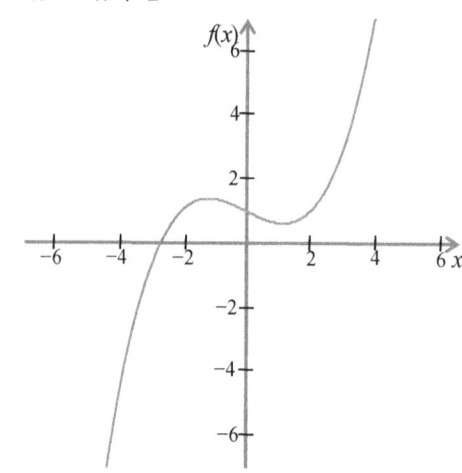

$$g(x) = \frac{1}{x+1}$$

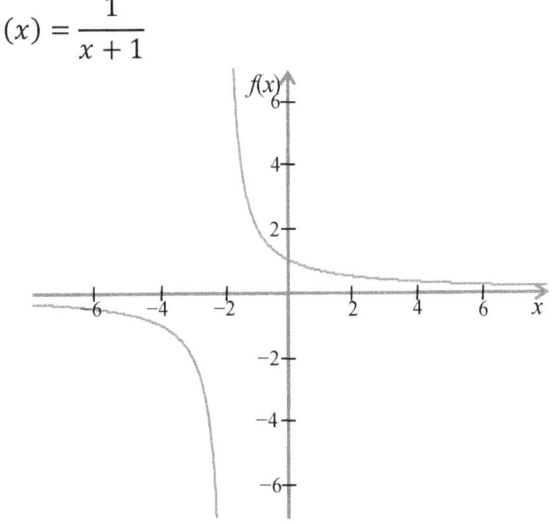

3. **Periodic Function**

A periodic function $f(x)$ with period a is a function such that $f(x) = f(x+a) = f(x+ka), \forall k \in \mathbb{Z}$.

The functions $f(x) = \sin x$ and $f(x) = \cos x$ shown above are examples of periodic functions. Their period is 2π.

The following is another example of a periodic function with period 2. The graph is in the range $-3 < x \le 4$.

$$f(x) = \begin{cases} 2x - 6, & \text{for } 3 < x \le 4 \\ f(x+2) = f(x) & \text{for all } x \end{cases}$$

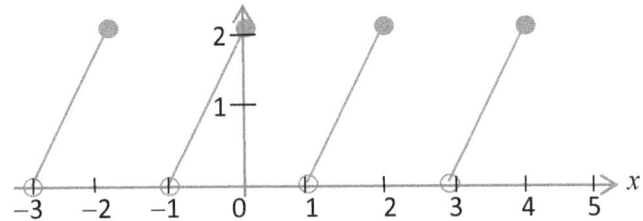

The basic range for the above function is $3 < x \le 4$. The graph is drawn by drawing the graph for the range

$3 < x \le 4$ and repeating it at intervals of 2 (the period) within the specified range.

The value $f(a)$ for any value a which is not within the range of the graph is given by

$f(a) = f(a - \text{closest lower multiple of the period to } a)$

For instance,

$$f(11.5) = f(11.5 - 2 \times 4) = f(3.5) = 1$$
$$f(-8) = f(-8 - 2 \times (-4)) = f(0) = 2$$

4. Compound Periodic Function

A compound periodic function is a function which is defined by two or more formulae in the basic range.

An example of a compound periodic function with period 5 is shown below. The graph is in the range $-1 < x \le 10$.

$$f(x) = \begin{cases} 2 - x, & \text{for } 0 < x \le 4 \\ 1 \text{ for } 4 < x \le 5 \\ f(x + 5) = f(x) \text{ for all } x \end{cases}$$

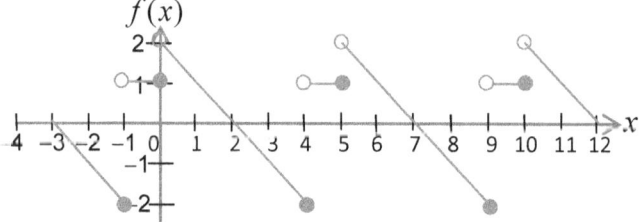

The value $f(a)$ for any value a which is not within the range of the graph is given by

$f(a) = f(a - \text{closest lower multiple of the period to } a)$

For instance,

$$f(23) = f(23 - 4 \times 5) = f(3) = -1$$
$$f(9.8) = f(9.8 - 5) = f(4.8) = 1$$

5. Continuous Function

A function $f(x)$ is continuous at $x = a$ if $\lim_{x \to a} f(x) = f(a)$

i.e. $f(x) = f(a)$ as $x \to a$ from above and from below.

The graph of a continuous function is unbroken. Examples of some continuous functions are: $f(x) = \sin x$, $f(x) = \ln x$, $f(x) = |x|$.

Below is the graph of $f(x) = \ln x$.

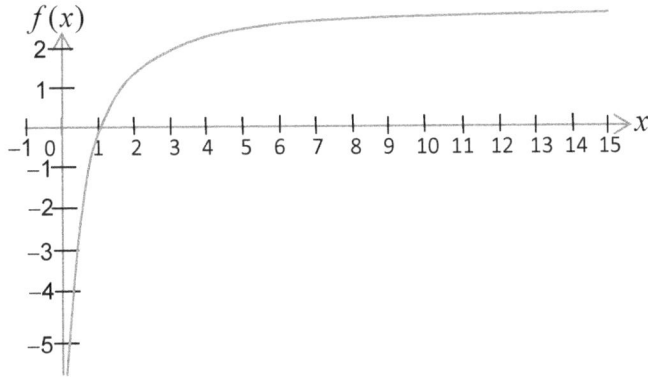

Example

Given that
$$f(x) = \begin{cases} -x^2 - x + 6, 0 \le x < 2 \\ kx - 4, 2 \le x \le 5. \end{cases}$$
Find the value of k for which the function $f(x)$ is continuous.

Solution

For continuity, $-x^2 - x + 6 = kx - 4$ when $x = 2$.
$$\Rightarrow -(2^2) - 2 + 6 = 2k - 4 \Rightarrow k = 2.$$

6. **Discontinuity**

If $f(x) \neq f(a)$ as $x \to a$ from above and from below, then $f(x)$ has a discontinuity at $x = a$.

E.g. the function $f(x) = \begin{cases} x - 1, x < 2 \\ x, x > 2 \end{cases}$, has a discontinuity at $x = 2$.

TOPIC 17
NUMERICAL METHODS

Solutions to Equations by Iterative processes

Many equations such as $x + \ln x = 1$, cannot be solved using the conventional methods. However, there are a handful of numerical methods which can be used to find approximate solutions to such equations.

Graphical Method of Solving Equations

Given the function $f(x) = 0$, we can rearrange this function as $g(x) = h(x)$. By drawing the graphs of $g(x)$ and $h(x)$, we can find the point(s) of intersection these graphs. The value(s) of x at the point(s) of intersection these graphs are the roots of the equation $f(x) = 0$. For following illustrate the required graphs for the given equations.

Equation	Required graph
$\ln x + x - 1$	$y = \ln x$ and $y = 1 - x$.
$\cos x - x + 1$	$y = \cos x$ and $y = x - 1$.
$e^{-x} - 3x + 3$	$y = e^{-x}$ and $y = 3x - 3$.

Example

Solve the equation $x - 1 + \sin x = 0$.

Solution

First rearrange the equation as $\sin x = 1 - x$. Then plot the graphs of $y = \sin x$ and $y = 1 - x$ on the same Cartesian axes. Where the two graphs intersect gives the solution of the equation $\sin x = 1 - x$.

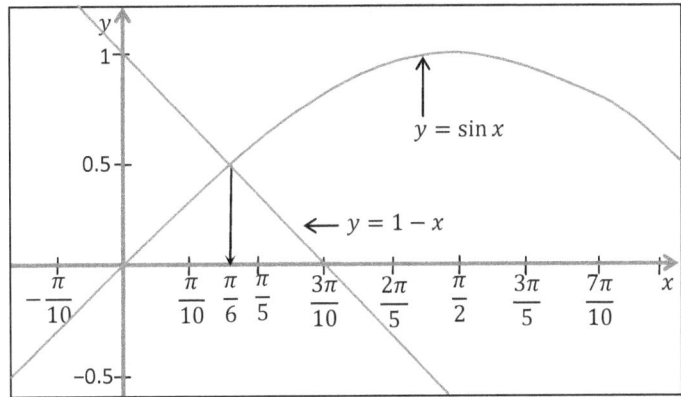

From the graph, $x \approx \dfrac{\pi}{6} \approx 0.52$.

Notice that it is easier to graph $y = \sin x$ and $y = 1 - x$ separately than graphing $y = x - 1 + \sin x$.

Change of Sign Method

If a function $f(x)$ is continuous in the interval $a < x < b$ and if $f(a)$ and $f(b)$ are of opposite sign, then there exist at least one real root of $f(x) = 0$ between a and b.

We can use this idea repeatedly to continuously narrow the interval to obtain the approximate root.

Interval Bisection or Midpoint Method

Interval Bisection or Midpoint Method is just an extension of the change of sign method. The procedure begins by finding the value of $f(x)$ for the extreme points of the given interval. By repeatedly halving the interval and finding the value of $f(x)$ for the midpoint, the side on which the root is can be found. This is done until the required degree of accuracy is attained.

Example

Show that a root α of the equation $e^x - 3x + 0.2 = 0$ lies between 0. 55 and 0.85 and use of the interval bisection (or midpoint method) to find α correct to 2 decimal places.

123

Solution

Let $f(x) = e^x - 3x + 0.2 = 0$

$f(0.55) = e^{0.55} - 3(0.55) + 0.2 = 0.28$

$f(0.85) = e^{0.85} - 3(0.85) + 0.2 = -0.01$

Therefore, a root of the equation lies between 0. 55 and 0.85

$$\text{Midpoint} = \frac{0.55 + 0.85}{2} = 0.7 \Longrightarrow x = 0.7$$

$f(0.7) = e^{0.7} - 3(0.7) + 0.2 = 0.11$

$$\text{Midpoint} = \frac{0.7 + 0.85}{2} = 0.78 \Longrightarrow x = 0.78$$

$f(0.78) = e^{0.78} - 3(0.78) + 0.2 = -0.56$

$$\text{Midpoint} = \frac{0.7 + 0.78}{2} = 0.74 \Longrightarrow x = 0.74$$

Therefore, $\alpha = 0.74$.

Method of Linear Interpolation or Chord Method

The method of linear interpolation states that If a function $f(x)$ is continuous in the interval $a < x < b$ and a root lies between this interval then a better approximation to the root of the equation $f(x) = 0$ will be given by

$$c = \frac{bf(a) - af(b)}{f(a) - f(b)}$$

Example

Show that $x^3 + 4x + 6 = 0$ has a root α in the interval -2 to -1 and hence find α correct to 1 decimal place by linear interpolation.

Solution

Let $f(x) = x^3 + 4x + 6$

$$\Rightarrow f(-2) = (-2)^3 + 4(-2) + 6 = -10$$

$$\Rightarrow f(-1) = (-1)^3 + 4(-1) + 6 = 1$$

Therefore, a root of the equation lies between -2 and -1.

$$c = \frac{bf(a) - af(b)}{f(a) - f(b)}, a = -2, b = -1$$

$$\Rightarrow c = \frac{-1(1) - (-10)(1)}{-10 - 1} = -\frac{9}{11} = 0.8 \Rightarrow \alpha = 0.8$$

The Newton-Raphson Method

The Newton-Raphson formula states that if a is an approximation to the root of the equation $f(x) = 0$, then a better approximation to the root of $f(x) = 0$ will be

$$b = a - \frac{f(a)}{f'(a)}.$$

Example

Use the Newton-Raphson method twice to find the root of the equation $x - 1 + \sin x = 0$, using 0.5 as a first approximation (x must be in radians for differentiation to be possible).

Solution

$f'(x) = 1 + \cos x$ and $a = 0.5$.

$$b_1 = a - \frac{f(a)}{f'(a)} = 0.5 - \frac{0.5 - 1 + \sin 0.5}{1 + \cos 0.5} = 0.49.$$

$$b_2 = b_1 - \frac{f(b_1)}{f'(b_1)} = 0.49 - \frac{0.49 - 1 + \sin 0.49}{1 + \cos 0.49} = 0.46.$$

Therefore, a better approximation to the root is $x = 0.46$.

TOPIC 18
MATRICES AND TRANSFORMATIONS

Definitions of a Matrix and Related Terms

A matrix is a rectangular array of numbers called **terms, elements** or **entries**.

$$\begin{pmatrix} -2 & 7 & 4 \\ 5 & 1 & 2 \\ 0 & 3 & 9 \end{pmatrix}$$

A horizontal line of elements of a matrix is called a **row**.

A vertical line of elements of a matrix is called a **column**.

A matrix with n rows and m columns has size or order $n \times m$ read "n by m" and is called an $n \times m$ matrix.

Types of Matrices

1. If $n \neq m$, the matrix is called a rectangular matrix.

2. If $n = m$, the matrix is called a square matrix.

$$\begin{pmatrix} 2 & 0 & -6 \\ 4 & 7 & 3 \end{pmatrix}$$

Rectangular matrix   Square matrix

3. A diagonal matrix is a square matrix in which all the entries not in the leading diagonal are zeros.

$$\begin{pmatrix} 5 & 0 \\ 0 & 3 \end{pmatrix}, \begin{pmatrix} 2 & 0 & 0 \\ 0 & 1 & 0 \\ 0 & 0 & 9 \end{pmatrix} \qquad \begin{pmatrix} 1 & 0 \\ 0 & 1 \end{pmatrix}, \begin{pmatrix} 1 & 0 & 0 \\ 0 & 1 & 0 \\ 0 & 0 & 1 \end{pmatrix}$$

Diagonal matrices   Unit or identity matrices

4. A unit or identity matrix is a square matrix with all its leading diagonal entries ones and the rest zeros.

5. A matrix whose size is $n \times 1$ is called a column matrix.

$$\begin{pmatrix} -6 \\ 2 \end{pmatrix}, \begin{pmatrix} 5 \\ 0 \\ 8 \end{pmatrix} \qquad (-6 \quad 2), \quad (5 \quad 0 \quad 8)$$

Column matrices   Row matrices

6. **A** matrix whose size is $1 \times n$ is called a row matrix.

7. A zero matrix is a matrix with all its entries zeros.

$$\begin{pmatrix} 0 & 0 \\ 0 & 0 \end{pmatrix}, \begin{pmatrix} 0 & 0 & 0 \\ 0 & 0 & 0 \\ 0 & 0 & 0 \end{pmatrix}, \begin{pmatrix} 0 \\ 0 \\ 0 \end{pmatrix} \qquad (3)$$

Zero matrices   Scalar

8. A 1×1 matrix is simply a single number and is called a scalar.

Matrix Algebra

1. **Addition and Subtraction of Matrices**

Addition and subtraction of matrices is possible only for matrices of the same size. Matrices of the same size are said to be compatible for matrix addition or subtraction. To add (or subtract) matrices, simply add (or subtract) corresponding entries.

2. **Scalar Multiplication of Matrices**

To multiply a matrix by a scalar, simply multiply each entry by the scalar.

3. **Equality of Matrices**

Two matrices are equal if their sizes are the same and their corresponding elements are equal.

4. **Multiplication of a Matrices**

Multiplication of two matrices **A** and **B** is possible only if the number of rows in the first matrix **A** is equal to the number of columns in the second matrix **B**. The size of the product of an $n \times m$ and an $m \times p$ matrix is an $n \times p$ matrix. Each row r, column c entry in the product is the result of the scalar product of the rth row of matrix **A** and the cth column of matrix **B**.

Note that addition of matrices is commutative but subtraction and multiplication are not.

Transpose of a matrix

The transpose of a matrix is the matrix obtained by interchanging the rows and columns of the matrix. e.g.

$$\mathbf{M} = \begin{pmatrix} a_1 & a_2 \\ b_1 & b_2 \end{pmatrix} \Rightarrow \mathbf{M}^T = \begin{pmatrix} a_1 & b_1 \\ a_2 & b_2 \end{pmatrix}$$

$$\mathbf{M} = \begin{pmatrix} a_1 & a_2 & a_3 \\ b_1 & b_2 & b_3 \\ c_1 & c_2 & c_3 \end{pmatrix} \Rightarrow \mathbf{M}^T = \begin{pmatrix} a_1 & b_1 & c_1 \\ a_2 & b_2 & c_2 \\ a_3 & b_3 & c_3 \end{pmatrix}$$

$$(\mathbf{AB})^T = \mathbf{B}^T \mathbf{A}^T$$

Determinant |M| of a 2 × 2 Matrix

$$\mathbf{M} = \begin{pmatrix} a_1 & a_2 \\ b_1 & b_2 \end{pmatrix} \Rightarrow |\mathbf{M}| = a_1 b_2 - a_2 b_1$$

The Determinant of a 3×3 Matrix

$$\mathbf{M} = \begin{pmatrix} a_1 & a_2 & a_3 \\ b_1 & b_2 & b_3 \\ c_1 & c_2 & c_3 \end{pmatrix}$$

$$\Rightarrow |\mathbf{M}| = \begin{vmatrix} a_1 & a_2 & a_3 \\ b_1 & b_2 & b_3 \\ c_1 & c_2 & c_3 \end{vmatrix} = a_1 \begin{vmatrix} b_2 & b_3 \\ c_2 & c_3 \end{vmatrix} - a_2 \begin{vmatrix} b_1 & b_3 \\ c_1 & c_3 \end{vmatrix} - a_3 \begin{vmatrix} b_1 & b_2 \\ c_1 & c_2 \end{vmatrix}$$

Properties of Determinants

1. If two rows (or two columns) of a square matrix **A** are identical, then det **A** $= 0$.
2. If all the entries in one row (or column) of a square matrix **A** are zero, $|\mathbf{A}| = 0$.
3. If two rows (or two columns) of a square matrix **A** are identical then $|\mathbf{A}| = 0$.
4. If two rows (or two columns) of a square matrix **A** are interchanged, then only the sign of $|\mathbf{A}|$ is changed.
5. The value of $|\mathbf{A}|$ is unchanged if a multiple of one row is added to or subtracted from another row, or if a multiple of one column is added or subtracted from to another column.
6. If one row (or column) of $|\mathbf{A}|$ is multiplied by $|\mathbf{A}|$, the resulting determinant is equal to $\lambda|\mathbf{A}|$. This implies that a common factor of the entries of one row (or column) is a factor of the determinant.
7. The value of a determinant is unaltered if the rows and columns are completely interchanged. This means that any property proven for rows is also valid for columns.
8. If **A** and **B** are square matrices of the same order, then $|\mathbf{AB}| = |\mathbf{A}||\mathbf{B}|$.
9. If **A** is invertible (i.e. **A** has an inverse), then $|\mathbf{A}^{-1}| = |\mathbf{A}|^{-1}$.
10. If **A** is an $n \times n$ matrix, then $|k\mathbf{A}| = kn\,|\mathbf{A}|$.

We can use the above properties to produce a matrix whose determinant is easier to evaluate and to simplify the evaluation of a determinant.

Area of a Triangle

Area of a triangle with vertices $(x_1, y_1), (x_2, y_2), (x_3, y_3)$ is given by $A = \dfrac{1}{2}\begin{vmatrix} 1 & 1 & 1 \\ x_1 & x_2 & x_3 \\ y_1 & y_2 & y_3 \end{vmatrix}$.

If the area of a triangle is zero, then the three points are collinear. Therefore, the condition that three points (x_1, y_1), (x_2, y_2), (x_3, y_3) are collinear is that $\begin{vmatrix} 1 & 1 & 1 \\ x_1 & x_2 & x_3 \\ y_1 & y_2 & y_3 \end{vmatrix} = 0$.

The Cofactors of the Elements of a 3×3 Matrix

The minor of an element of a 3×3 matrix is the determinant of the 2×2 matrix left after crossing the row and column of the element. Let $A = \begin{pmatrix} a_1 & a_2 & a_3 \\ b_1 & b_2 & b_3 \\ c_1 & c_2 & c_3 \end{pmatrix}$. The minor of b_3 is obtained by crossing row 2 and column 3 as in (i) below then finding the determinant of the 2×2 matrix $\begin{pmatrix} a_1 & a_2 \\ c_1 & c_2 \end{pmatrix}$ left.

$$\begin{pmatrix} a_1 & a_2 & a_3 \\ b_1 & b_2 & b_3 \\ c_1 & c_2 & c_3 \end{pmatrix}$$

$$\begin{pmatrix} + & - & + \\ - & + & - \\ + & - & + \end{pmatrix}$$

(i) (ii)

The **associated signs** of the minors of a 3×3 matrix are as shown in (ii) above.

The **cofactor of an element** is a combination of the minor and its associated sign.

The Adjoint (or Adjugate) of a 3×3 Matrix

The adjoint of a 3×3 matrix **A** denoted by AdjA is obtained by replacing the entries of the transpose matrix by their cofactors.

The Inverse A^{-1} of a Square Matrix A

The Inverse A^{-1} of a Square Matrix **A**, defined as

$$A^{-1} = \frac{1}{|A|} \times \text{AdjA}.$$

System of Equations

A system of simultaneous equations is a set of two or more equations in two or more unknowns. A solution of such a system is a set of values of the unknowns that satisfies every equation of the set.

Consider the equations

$$L_1 : a_1x + b_1y + c_1 = 0, \quad L_2 : a_2x + b_2y + c_2 = 0$$

1. L_1 Intersects L_2 $\Longleftrightarrow$ the equations have a unique solution.

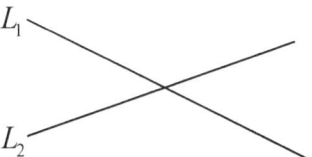

2. L_1 is parallel to L_2 $\Longleftrightarrow$ the equations are inconsistent or have no solution.

$$L_1 \underline{\hspace{4cm}}$$

$$L_2 \underline{\hspace{4cm}}$$

3. L_1 is a scalar multiply of L_2 $\Longleftrightarrow$ the equations have an infinite set of solutions. Such equations are said to be linearly dependent.

$$L_1 \underline{\hspace{4cm}} L_2$$

Conditions for three Lines to be Concurrent

$$\left. \begin{array}{l} L_1 : a_1x + b_1y + c_1 = 0 \\ L_2 : a_2x + b_2y + c_2 = 0 \\ L_2 : a_2x + b_2y + c_2 = 0 \end{array} \right\} \text{ are concurrent} \Rightarrow \begin{vmatrix} a_1 & b_1 & c_1 \\ a_2 & b_2 & c_2 \\ a_3 & b_3 & c_3 \end{vmatrix} = 0$$

Three Variables

Consider the three planes

$$\pi_1 : a_1 x + b_1 y + c_1 z = d_1$$
$$\pi_2 : a_2 x + b_2 y + c_2 z = d_2$$
$$\pi_3 : a_3 x + b_3 y + c_3 z = d_3$$

1. π_1, π_2 and π_3 intersect at one point $\Rightarrow$ unique solution.

2. π_1, π_2 and π_3 are identical or intersect in a line $\Rightarrow$ The equations have an infinite set of solutions and the equations are linearly dependent.

3. The set of equations have no solution and are inconsistent if:

    (i) π_1, π_2 and π_3 are all parallel or

    (ii) Any two of the planes are parallel but distinct

    (iii) Any of the planes is parallel to the line of intersection of the other two planes.

Matrix transformation in 2-dimensions

1. The image (x_1, y_1) of the point (x, y) under a transformation represented by the 2×2 matrix $M = \begin{pmatrix} a & b \\ c & d \end{pmatrix}$ is given by $\begin{pmatrix} x_1 \\ y_1 \end{pmatrix} = \begin{pmatrix} a & b \\ c & d \end{pmatrix}\begin{pmatrix} x \\ y \end{pmatrix}$.

2. If during such a transformation, the area changes, then

    Image area=Object area $\times$ |**M**|

This means that |**M**| is the scale factor of the transformation. If $|\mathbf{M}| = 0$, we call the matrix a singular matrix. A singular matrix collapses two dimensional space to a point or a line passing through the origin.

3. To find the transformation matrix $\begin{pmatrix} a & b \\ c & d \end{pmatrix}$ simply find the images $\begin{pmatrix} a \\ c \end{pmatrix}$ and $\begin{pmatrix} b \\ d \end{pmatrix}$ of the unit base vectors $\begin{pmatrix} 1 \\ 0 \end{pmatrix}$ and $\begin{pmatrix} 0 \\ 1 \end{pmatrix}$ and place them as columns of the transformation matrix.

Matrix transformation in 3-dimensions

1. The image (x_1, y_1, z_1) of the point (x, y, z) under a transformation represented by the 3×3 matrix

$$\mathbf{M} = \begin{pmatrix} a_1 & b_1 & c_1 \\ a_2 & b_2 & c_2 \\ a_3 & b_3 & c_3 \end{pmatrix} \text{ is given by } \begin{pmatrix} x_1 \\ y_1 \\ z_1 \end{pmatrix} = \begin{pmatrix} a_1 & b_1 & c_1 \\ a_2 & b_2 & c_2 \\ a_3 & b_3 & c_3 \end{pmatrix} \begin{pmatrix} x \\ y \\ z \end{pmatrix}$$

2. If during such a transformation, the volume changes, then

Image Volume =Object Volume $\times |\mathbf{M}|$

This means that $|\mathbf{M}|$ is the scale factor of the transformation.

If $|\mathbf{M}| = 0$, then the matrix is said to be singular and collapses three dimensional space to a line or plane passing through the origin.

3. To find the transformation matrix $\begin{pmatrix} a_1 & b_1 & c_1 \\ a_2 & b_2 & c_2 \\ a_3 & b_3 & c_3 \end{pmatrix}$ simply find the images $\begin{pmatrix} a_1 \\ a_2 \\ a_3 \end{pmatrix}$, $\begin{pmatrix} b_1 \\ b_2 \\ b_3 \end{pmatrix}$ and $\begin{pmatrix} c_1 \\ c_2 \\ c_3 \end{pmatrix}$ of the unit base vectors $\begin{pmatrix} 1 \\ 0 \\ 0 \end{pmatrix}$, $\begin{pmatrix} 0 \\ 1 \\ 0 \end{pmatrix}$ and $\begin{pmatrix} 0 \\ 0 \\ 1 \end{pmatrix}$ and place them as columns of the transformation matrix.

Invariant points and lines

For any transformation matrix of the form $\begin{pmatrix} a & b \\ c & d \end{pmatrix}$, the origin is invariant. Some transformations such as shears have lines that are invariant.

Invertible Matrices

If a square matrix **A** has an inverse $\mathbf{A}^{-1}$, it is said to be **invertible**. If that is the case then $\mathbf{AA}^{-1} = \mathbf{A}^{-1}\mathbf{A} = \mathbf{I}$.

If **A** and **B** are square matrices of the same size, both having inverses $\mathbf{A}^{-1}$ and $\mathbf{B}^{-1}$ then $(\mathbf{AB})^{-1} = \mathbf{B}^{-1}\mathbf{A}^{-1}$.

If a transformation represented by a matrix **M** maps an object into an image, then the inverse $\mathbf{M}^{-1}$ maps the image into the object.

Compound Transformations

The matrix operator of the compound transformation **A** followed by **B** is **BA**. The order in which the transformations are performed is very important and must be stated.

www.ingramcontent.com/pod-product-compliance
Lightning Source LLC
Chambersburg PA
CBHW071444180526
45170CB00001B/458